HOME
ENERGY
HOW-TO

A Popular Science Book

HOME ENERGY HOW-TO

by A.J. Hand

Drawings by
Carl J. De Groote

POPULAR SCIENCE

HARPER & ROW
New York, Evanston, San Francisco, London

Library of Congress Catalog Card Number: 76-053195
ISBN: 0-06-011774-5

To my father

CONTENTS

WHAT THIS BOOK CAN DO FOR YOU

Energy is like the weather. Everybody talks about it, but nobody does anything about it. Hardly a day goes by when you don't hear about the energy shortage: People complaining about their electric bills or the high cost of heating their homes. Newsmen covering the latest meeting of OPEC. Neighbors complaining about the Arabs. But that's about as far as it goes. Very little action is taken.

Successive government administrations have sat on their hands, looking the other way, and hoping the whole problem would pass or that it would at least remain tolerable until the next administration's takeover. The United States—the world's largest consumer of energy by far—continued into the late 1970s the only major country in the world without a comprehensive energy policy. People who have wanted action on energy have had to become do-it-yourselfers.

That's what this book is all about: Ways you can save energy. Why bother? Three good reasons: Save energy and you save money; that's an idea that should appeal to anyone. Save energy and you cut pollution. Finally, save energy and you conserve a natural resource that is, in most cases, nonrenewable. When you burn a gallon of oil, a cubic foot of gas, or a pound of coal, one thing is certain: That bit of fuel is gone forever.

Why should you care? The simple fact is these fuels are all finite resources. Someday they will run out. Sure, we have enough coal to last for a few hundred years. But petroleum is going fast. It may not dry up in your lifetime, but it may in the lifetime of our children—or the lifetime of their children. When that happens, what will warm their homes? What will drive their transportation systems and their economy? The power of nuclear fission? Not likely. Unless science comes up with some safe and sure way of handling radioactive nuclear wastes, it's more likely that fission will poison people rather than serve them.

Radioactive wastes must be isolated from the environment for a period of 200,000 years before they become safe. Who is going to watch over them for that length of time? If radioactive wastes had been present in the year 1 A.D., today they would be only 2,000 years old. They would have progressed only one percent of their way toward harmlessness. Think of the earthquakes, volcanic eruptions, revolutions, and world wars that have taken place over the past 2,000 years. Multiply them by 100. Meanwhile, who's minding the radioactive store?

What about nuclear fusion, you say? That's a much cleaner process than fission. True, but it also involves temperatures so high they'll melt any material known to man. How do you harness temperatures like that? Right now science is tinkering with the idea of containing the reaction in a magnetic field. Still, nobody is certain that fusion will be harnessed in the foreseeable future, if at all.

I could go on, but it's clear enough. Energy is worth saving. And you'll have to do it yourself. What can you do in your home? What will this cost you? And how much will it save? This book answers these questions. Let's take a quick trip through the book's twelve chapters.

To begin, we'll examine energy and your living habits. You'll learn how to save energy just by making simple changes in your lifestyle. In most cases the changes will cost you nothing, and the savings can be significant. For example, you can lower your thermostat and wear a sweater. As a result, home heating costs can drop as much as 35 percent.

Next comes a chapter on insulating your home. You'll see which jobs are most important. You'll learn how to do them, how much they'll cost, and how much they'll save you. Insulating your attic—always your best insulation investment—can save you hundreds of dollars a year for the rest of your life. This may be a simple half-day job in many cases.

In the chapter on caulking and weatherstripping, you'll find some more jobs you can do yourself in a day. Detailed instructions show you how. When you're done you can expect to recover your costs in a year and reap an annual savings as high as $125 from then on.

Next we'll cover windows and doors, the two biggest energy-wasting surfaces in your home. Read this chapter and you'll learn how to cut heat losses dramatically. You'll also learn how to turn your windows into energy collectors.

After windows and doors come two chapters, one on conventional heating systems, the other on home cooling systems. Both tell you how to get the most from your energy dollar. You'll learn how to tune your present system and modify it for maximum effciency. And you'll find important tips on buying new heating and cooling systems.

Next we'll take a look at energy-efficient architecture. You'll learn how to build your next home—or modify your present one—to save energy in dozens of ways. We'll cover the size, shape, and orientation of your home; ways to collect free energy; ways to landscape for better heating efficiency and better cooling in the summer.

This far, the book concentrates essentially on saving energy through living habits and containing energy through home modifications. The rest of the book describes the ways you can produce and harness alternate forms of energy. Here we'll explore all the possibilities for tapping the resources of sun, wind, water, and biofuels as well as for heating with wood.

In the chapter on energy from the sun, we'll examine the most practical solar applications for homes in both northern and southern climates. You'll see how the basic air-cooled and liquid-cooled collectors work, and you'll learn how to determine which would be better suited for your use. You'll see how to size and buy commercially produced systems and also how to make your own from scratch.

Maybe the answer to your energy needs is blowing in the wind. If so, the chapter on wind power will tell you how to capture it. You'll learn what wind generating systems cost, what they can do for you, and where you can get them. Water pumping mills are covered, too.

The seemingly outmoded practice of heating with wood has probably become the most popular, most effective, and least expensive alternate-energy system available today. I'm presently supplementing my home heating system with a small wood-stove and I've cut my fuel bills about 60 percent as a result. The chapter on wood heating will show you how you can do the same.

Water power is free, and it's the best way to generate your own electricity if you have access to a decent stream. The chapter on water power will tell you how to measure your stream's potential and how to put that potential to use.

Finally, we finish up with a chapter on the biofuels. Here you'll find suggestions for turning organic wastes into methane gas and fertilizer, as well as information on running your car on wood.

That's it—a dozen chapters on saving energy and producing your own. Not all the ideas and information presented will apply to your situation. But those that do should enable you to cut your energy costs to a small fraction of what they've been. At the same time they'll help you cut down on the pollution you create whenever you use energy. And they'll help conserve the energy sources for generations to come.

A. J. Hand

PART ONE

ENERGY AND YOUR LIVING HABITS

No discussion of energy conservation in the home could be complete without at least brief mention of personal living habits. Over the years we've become so accustomed to cheap energy that we often waste energy without realizing what we're doing. Or without realizing that the waste isn't necessary.

Probably the simplest and most effective method of cutting energy use in the home is to learn a little thermostat discipline. It sometimes seems that we become intoxicated with the power we have over our home environment. If we feel chilly during the heating season, we exercise our power by turning up the thermostat. If we feel warm in the summer, we turn on the airconditioner. Or, if that's already running, we turn down the thermostat. In many cases a simple change of clothing would achieve the same effect.

I've visited a couple in winter who greeted me at the door in stocking feet or light slippers. The husband wears a light cotton shirt with short sleeves, the wife a light sleeveless house dress. A quick glance at the liquid crystal thermometer in the den reveals a temperature of 78 degrees.

Oddly enough, that 78-degree temperature—so comfortable to them in the winter—is much too hot in the summer. But no problem, the central airconditioner takes care of that.

Maybe I keep my house at the opposite end of the temperature extreme. Some of my acquaintances certainly think so. During the winter I keep the house around 62 to 64 degrees. At first this seemed pretty cold. But a heavy sweater and a little adaptation on my part soon made the situation bearable, even comfortable. Besides the sweater, there's another trick that makes 62 degrees bearable. Most of my heat comes from a woodstove. Whenever the stove can't keep the house warm enough, the oil burner in the basement cuts in. I've located the woodstove in the living room where its radiant heat can keep me warm even when I'm relaxing in a chair—watching television or reading —just a few feet from the stove. I know from experience that it's much harder to feel warm at 62 degrees when I don't have this little pocket of radiant energy to retire to.

On the other hand, I spend a lot of time in my workshop during the winter. The temperature in the shop is around 55, but it's warm enough for comfort because my physical activity generates body heat to offset the cold. My shop is no place to relax and read a book, but it's just about right for cutting wood or driving nails.

Is all this "cold" bad for my health? I can't prove it, but it seems to my wife and me that we have fewer problems with head colds now than we did a few years ago when we kept the temperature up around 72 degrees all winter. How much fuel has this lowering of the thermostat saved? The first winter we cut our consumption

by about one-third over the previous winter, simply by lowering the thermostat 10 degrees. The second winter, using a tiny Jotul woodstove, we cut our consumption roughly in half again. So the combination of stove plus low thermostat setting has cut our oil consumption about two-thirds.

Thermostat discipline may be the best energy-saving habit you can develop, but there are other ways to save energy through discipline. Most obvious is to discipline yourself into a reflex habit of turning off lights that aren't being used. Incidentally, fluorescent lighting uses much less electricity than incandescent bulbs. A 40-watt fluorescent gives about the same amount of light as a 150-watt bulb. But there's a hitch here: Fluorescents use a lot of current getting started, and frequent switching on and off will shorten fluorescent lamp life. So there's a tradeoff. It's not smart to turn off a fluorescent if you know you'll be needing it again in an hour or less. You won't save electricity by doing so, and you'll shorten lamp life too.

Window, door, and curtain discipline can help you save energy, too. Learn habits that will help you make the most of all the forms of energy available to you. During the day, for example, open curtains and shades on all windows that face the sun. Allow them to collect solar energy whenever it is available. When it isn't, close curtains and shades to retain heat. Do the same with doors. I've discovered I can collect solar heat by

Foot-pedal control of kitchen faucets can save a lot of hot water, such as when rinsing dishes. *Courtesy T & S Brass and Bronze Works.*

opening my front door (but leaving the storm door closed) during the morning hours. The sun shines in through the storm door and hits the dark tiled floor inside. Soon the floor is warm—almost too hot to touch. But as soon as the sun swings around to where it can't shine in through the storm door, I close the wooden door to prevent heat loss.

Strange as it seems, many people don't appreciate the fact that wasted hot water is wasted heat. Any time you can save hot water you can save energy. That's why washing laundry in cold water makes sense. As do quick showers, rather than long showers or baths. Foot pedal controls for the kitchen faucet are another way to save hot water; with foot control you can run the water only when you

need it. Long ago Bucky Fuller recognized that rinsing and washing require much less water if very fine sprays are used. Fine-spray heads for showers and sinks are available to help you make use of this principle.

Other good hot water habits to develop:

- Drain a couple gallons of water from your hot water tank twice a year. This will get rid of sediment that insulates the water from the flame (gas water heater) that heats it.
- Set the thermostat to 120 if you have no dishwasher. Set it to 140 if you have a dishwasher. Low settings prolong heater life, too.
- Turn the heater thermostat all the way down when you go on vacation.
- Fix dripping faucets.
- Always wash full loads in clothes and dishwashers.

Energy-saving habits like these can be applied to almost every phase of your living that involves energy. A complete list of good habits would be virtually endless, but here are some more ideas that should help you get you thinking along energy conserving lines:

- If you have a fireplace, keep the damper closed during the winter when no fire is burning. Use a snuffer (sheet metal box to cover the fire) to kill a fire shortly before you retire. Then close the damper when the fire is dead out. Alterna-

To achieve the 75 percent efficiency potential of smooth-top ranges, you need cookware that sits flat on heating elements. *Courtesy Corning Glass Works.*

tively, install tight-fitting glass doors over the fireplace opening.
- Leaving your damper open in summer can help exhaust hot air from your home. This works best at night.
- When cooking, heat the least amount of water required to do the job. Want a cup of coffee? Heat a cup of water. Once it boils, turn off the heat. When boiling foods, turn down the heat once the water reaches a boil.
- Dry clothes on a line instead of in a

dryer whenever possible. If you use a dryer, don't overdry. Use the automatic dryness-sensing control if the dryer has one.
- On electric ranges, use flat-bottomed pans that match the size of the burner element. Aluminum or copper-bottomed pans will give up to 80-percent efficiency on electric stoves.
- Smooth-top ranges made for ceramic cooking utensils will give about 75 percent heating efficiency, but this drops to about 35 percent with metal pans. Use the right utensils.
- Turn off the range just before foods are done. Let the residual heat in the pan and the stove element finish the job.
- With gas stoves, use pans large enough to cover the flame. Dark colored pans are best.
- Have a serviceman adjust the flames to burn clean and blue and set the pilots as small as they will go while still providing reliable lights.
- When you use the oven, try to cook more than one item at a time, even if this means you're baking something you won't eat until a later meal. Turn off the oven about 10 minutes before the food is done and let the food coast on stored energy in the oven.
- Combination dishes all cooked in the oven or in the same pot on one burner save energy.
- Use special-purpose cooking appliances (waffle irons, toaster ovens,

electric skillets, etc.) if you have them. They are usually more efficient than a range. On the other hand, it doesn't pay to buy one of these appliances just to save energy unless you use it quite often.

- Black-and-white TV uses about half the energy of color TV. Solid state circuitry saves money, too. Instant-on TV uses electricity even when it's turned off. Unplug it or put a switch in the power cord so you can really turn the set off.
- Open refrigerator and freezer doors as little as possible. Choose manual defrost models. Turn off the door heating elements (anti-sweating device) if your refrigerator has a switch, by turning it to the "dry" setting.

Okay, those are just a few ways you can save energy with good living habits. Is all this endless niggling and Btu pinching worth the effort? Yes. One study of energy use compared two identical New Jersey houses, side by side, occupied by families of similar size. The amount of gas used to heat each house was recorded. Consumption differed by 50 percent. Living habits make the difference.

INSULATION

From an energy standpoint, insulation is the most important part of your home. Once installed, it does its job forever, without maintenance, year-round.

KEEPING HEAT WHERE YOU WANT IT

The basic secret to cutting your heating and cooling costs is keeping heat where you want it. In winter, for example, you put heat into your home by burning fuel. Now your goal is to keep it there. If your home were perfectly insulated and sealed, theoretically the heat would last all winter — actually, forever. Heat doesn't disappear or wear out. It isn't consumed. It simply escapes from your home. Dis-

counting the need for fresh breathable air, if you could prevent this escape you wouldn't need a furnace, except for the initial heating. And conversely, in the summer, if you could prevent outside heat from getting into your home, you'd never need an airconditioner.

Of course perfectly insulated homes don't exist, but they are a goal to work toward. In the past, fuel to heat our homes was so cheap that we'd build our houses with little or no provision for energy conservation. Why bother with the expense of insulation and weatherstripping when it might take 15 or 20 years to pay for itself? Today that kind of thinking is obsolete. When the price of fuel

SAVE IN HEATING

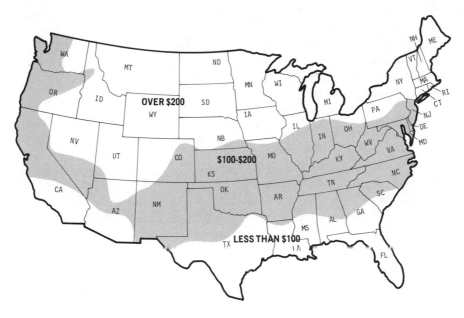

SAVE IN COOLING

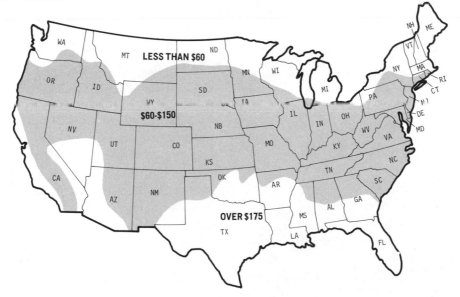

The importance of attic insulation is summarized in these diagrams. Calculations are based on a 1,000-square-foot uninsulated attic to which R-19 insulation is added.

doubles, energy-conserving measures pay off in half the time. And they make more than twice as much sense. In this age when fuel rationing can become a reality overnight, good insulation and sealing can spell the difference between a home that's warm all winter and one that runs out of fuel in February. Energy conservation becomes more than a simple matter of cost vs. savings when it turns out to be the only way to stay warm. What dollar value do you place on insulation and sealing when they're the only things that stand between you and an indoor ice age?

THERMAL HOME IMPROVEMENTS:

COSTS, SAVINGS, PAYOFF TIMES

IMPROVEMENT	COST	YEARLY SAVINGS	YEARS TO PAY OFF
Caulk and weatherstrip doors and windows	$76.30	$88.92	.85
Install glass storm windows	$120	$44.56	2.7
Insulate attic to R-22 with mineral wool	$216	$283.80	.76
Insulate basement walls two feet below ground to R-11 with mineral wool	$210	$36.20	5.8
Insulate frame walls with UF foam (contractor applied)	$960	$138	7

The table above shows savings resulting from insulation improvements for a house in Chicago. It has 1,200 square feet of living space, 12 windows, two doors, a full basement, oil heat, and central airconditioning. Because base prices for fuel oil, electricity, and insulation materials tend to increase at similar rates from year to year, the relationships of costs to yearly savings, and thus payoff times remain relatively stable. Thus, you might be able to make some estimations on savings potentials for your home based on factors noted in this table.

INSULATION

Insulation will save you more fuel, money, and discomfort than any other improvement you can make on your home, so let's deal with it first. Any discussion of insulation has to answer three main questions: Where do I put it? How much do I need? What kind should I use?

Where to Put Insulation. Quite simply, you should put insulation between any warm inside area and any cold area, inside or out. Think of insulation as a blanket holding heat in, separating it from any place that's cold. This includes unheated basements and garages as well as the outdoors. Of course, if you have a limited amount of money to spend on insulation you should start by putting it where it will do the most good. This is almost always in the attic. Luckily, this is almost always the easiest part of your home to insulate. You should be able to handle the job in a few hours if the attic is unfinished and unfloored. The more finished the attic, the harder it will be to insulate.

After you've insulated your attic, you have a couple of choices. You can insulate your house walls or your basement walls or ceiling. Insulating the exterior walls of a finished home is a job for a contractor—you can't do it yourself. So your costs for that job will be higher than for do-it-yourself insulating of the basement wall or ceiling. On the other hand, insulating your walls will save more money in

INSULATION AND HEAT FLOW: TERMS YOU MAY FIND HELPFUL

Btu. British thermal unit. The Btu is a measure of heat, specifically, the amount of heat required to raise one pound of water one degree F. Heat flow is often expressed in Btu per hour or Btuh.

Conduction. This is the flow of heat through an object. Some materials are good conductors, others are bad. Bad conductors make good insulators, and vice versa. Heat flow by conduction is always from the hot side to the cold side, and the flow can be in any direction; up down, left or right.

Convection. This is the flow of heat through a liquid or gas, and it is accomplished by actual movement of the liquid or gas. In convection, the direction of flow is upwards. For example, hot air, being less dense than colder air, rises. This sets up circulation (convection currents) that carry heat from a hot surface to a cold surface.

Radiation. Any object that is warmer than its surroundings will radiate heat in the form of waves. The heat you feel when you stand in the sun is the sun's radiant heat.

U. The overall heat transmission coefficient. This is the amount of heat transmitted in one hour through a square foot of a building section for each degree F of temperature difference between the air on the cold side and on the warm side of the section. U is expressed in Btuh.

R. This is the resistance to heat flow or the reciprocal of U, sometimes expressed as 1/U. Insulation is often labeled in R values. Example: common $3\frac{1}{2}$-inch mineral wool batts are rated at R-11. R values are additive. Double the thickness of insulation and you double its R. Or add the R values for the insulation, sheathing, siding, and air films in a wall and you have the R value of the entire wall section.

C. A unit of thermal conductance. The amount of heat per hour (Btuh) transmitted from surface to surface of one square foot of material or a combination of materials for each degree F of temperature difference between the two surfaces. Note that this value is not expressed in terms of "per inch of thickness," but from surface to surface no matter what the thickness.

k. The amount of heat (Btu) transmitted in one hour through one square foot of a homogeneous material one inch thick for each degree F of temperature difference between the two surfaces of the material. Good insulation has a low k.

1/k. The resistance to heat flow of one inch of a homogeneous material. Measured from surface to surface.

x/k. The resistance to heat flow of x inches of a homogeneous material. Measured from surface to surface.

f. Film or surface conductance. Btuh transmitted from one square foot of a surface to the air surrounding the surface for each degree F temperature difference. The symbols f_i and f_o designate inside and outside conductances.

1/f. Film or surface resistance. The resistance to heat flow of an air film adjacent to a surface. $1/f_i$ designates resistance of an inside surface resistance; $1/f_o$ is the resistance of an outer air film.

a. Thermal conductance of an air space. The Btuh transmitted across an air space on one square foot for each degree F temperature difference.

1/a. Air space resistance to heat flow.

the long run than either of the basement jobs. If your basement is heated, you may elect to insulate the walls to a level about two feet below the level of the ground outside. This is a fairly easy job, and the cost normally runs only about $200 to $300, including the cost of a finished wall over the insulation. If your home is built over a weathertight crawl space, you may want to insulate the crawl space walls. This is cheaper and easier than insulating basement walls because finished appearance is not important. If your home sits over an unsealed crawl space (on piers for example) or if a portion of it is over an unheated garage or porch, you may want to insulate the floor of the heated area. Savings here can be quite large, and costs are low. So if your home is in this category, floor insulation is a good second choice after your attic is done.

How Much Insulation? All insulation is rated according to its ability to resist heat flow. The rating, called the R-value, is usually marked on the insulation package when you buy mineral wool blankets or batts. But other forms of insulation are unmarked, so it pays to know the ratings of various types. See the accompanying table, "How Thick Should Your Insulation Be?"

Different parts of your house require different R-values, so study the map, "Which R-values Go Where?" to make sure you install at least the minimum amount of insulation recommended for each application.

What Kind of Insulation? This depends on a number of factors. So let's cover the various types of insulation and discuss the advantages and disadvantages of each.

Urethane foam. This plastic foam has the highest insulating value of any material currently in use. An inch of urethane is worth two inches of mineral wool, so it's most useful where you need a high R-value but have little space for insulation. But it presents a serious drawback: *It burns and releases toxic fumes in the process.* Whenever it is used it should be covered with ½-inch gypsum wallboard to assure fire safety.

Ureaformaldehyde (U-F) foam. This is the newest and in many ways the best new insulating material on the market. It ranks right behind urethane in R-value, and it's fire resistant. In fact, fire department officials in Minneapolis credit U-F foam with stopping a garage fire from reaching into the attached house. U-F is applied in liquid form—shot from a gun—so it can be squirted into fin-

HOW THICK SHOULD YOUR INSULATION BE?

	BATTS OR BLANKETS		LOOSE FILL (POURED IN)			RIGID PLASTIC FOAMS		
	glass fiber	rock wool	glass fiber	rock wool	cellulosic fiber	urethane	U-F	styrene
R-11	3½"-4"	3"	5"	4"	3"	1½"	2"	2¼"
R-19	6"-6½"	5¼"	8"-9"	6"-7"	5"	2¾"	3¾"	4¼"
R-22	6½"	6"	10"	7"-8"	6"	3"	4"	4½"
R-30	9½"-10½"	9"	13"-14"	10"-11"	8"	4½"	5¾"	6½"
R-38	12"-13"	10½"	17"-18"	13"-14"	10"-11"	5½"	7½"	8½"

WHICH R-VALUES GO WHERE?

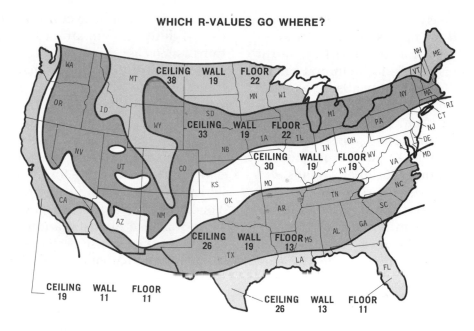

CEILING 38 WALL 19 FLOOR 22

CEILING 33 WALL 19 FLOOR 22

CEILING 30 WALL 19 FLOOR 19

CEILING 26 WALL 19 FLOOR 13

CEILING 19 WALL 11 FLOOR 11

CEILING 26 WALL 13 FLOOR 11

These R-values are currently recommended by Owens-Corning for achieving highest net savings on heating and cooling costs. As fuel costs rise, so will the optimum R-values. Insulation considered overadequate today may be just right a few years from now.

ished walls, floors, ceilings, all places you can't reach with many other forms of insulation. It would definitely be my first choice for the insulation of existing frame walls. The only drawback is the price, since a contractor must shoot it in. Even so, a typical frame-wall installation will pay for itself in four to eight years.

Polystyrene foam (Styrofoam). This material has the same shortcoming as urethane: It lacks fire resistance. Its greatest potential is in new construction where it can be used in tongue-and-groove board form, replacing standard wooden sheathing. In this application it can be combined with ordinary mineral wool between the wall studs to provide about double the insulation value of the mineral wool alone. This is ideal for solar homes, and it raises the cost of a typical home only about $135. Urethane sheathing boards are also available. Ordinary foam board not designed for sheathing can be used to insulate a finished home from inside, but it must be covered with ½-inch gypsum board, and it will cause all kinds of trim problems around windows, doors, and electrical outlets.

Mineral wool. This material is pretty much the standard material of today. It's cheap, comes in a variety of forms to fit most situations, and it does a fairly good job. Blankets and batts are easy to place between studs and joists or rafters. Mineral wool is available with a vapor barrier, and it is fire resistant. In loose form it can be blown into existing walls by contractors although it has a tendency to settle after a time, leaving uninsulated gaps. One problem to consider is that R-values that should be considered minimum for best dollar savings will continue to rise in relation to rising fuel prices. But there won't be room for extra-thick mineral wool in many applications, because this material completely fills the cavities in stud walls, and in between rafters and joists in many homes. Here, you'll need insulating material with a lower k-value (that which allows less heat to escape for a given thickness).

Cellulosic fiber. This fibrous material is used as a poured or blown-in loose-fill insulation. Due to its smaller tuft size it tends to distribute better than loose mineral wool. It also has a slightly higher insulation value. Drawbacks: It isn't vermin or fire proof. Some types are chemically treated to resist fire, but there is some question as to the effectiveness of the treatment after prolonged exposure in a hot attic. If you use this material in your home, check the bags to see that the fiber meets federal specifications.

Vermiculite and perlite. These are lightweight forms of mineral insulation that pour and distribute well in tight spaces. Both are expensive, and neither has a very good k rating. Both types are best used for small areas where access is difficult. Example:

Pour them down between wall studs; they'll filter down around obstructions such as pipes and electrical cables and outlets better than other types of loose fill.

Whenever you have a choice of insulation materials, it usually makes most sense to choose the one that gives the most R-value for the money. This isn't always the case, however. You might be willing to pay a little more for ease of installation, fire resistance, or vermin resistance. Just don't forget that one insulation will work as well as the next as long as R-ratings are the same. R-11 urethane won't stop heat flow any better than chopped newspaper rated at R-11.

Vapor Barriers. Warm air inside your home contains a lot of moisture. This moisture can condense on any cold surface it reaches. You've probably seen condensation on the inside of your windows on a cold winter day. Well, the same thing can happen in your walls, or under your roof. Here's why. When you insulate a wall or a roof, you prevent heat from inside your home from reaching wall or roof sheathing. As a result these layers are almost as cold as the outside air. When moist air from inside the house touches the cold sheathing, condensation can result. It can soak the sheathing and cause rot. And it can soak the insulation and cut its effectiveness.

So that's where the vapor barrier comes in. It blocks the flow of water vapor before it can reach the cold surface. It is always applied to the warm side of the insulation—toward the side that is heated in winter. If you put it on the cold side by mistake, the barrier itself will become cold, and water will condense right on the barrier. Blankets and batts usually come with a paper or aluminum foil vapor barrier already attached. The foil barrier has an advantage; it reflects heat back into the living space, raising the R-value of an air space within a wall from 1 to 3½. But for the foil to be effective, the air space must be at least ¾-inch thick.

If you are using an insulation material without a vapor barrier, in most cases you should add one. Plastic film is the cheapest and easiest to add. More about that later. Both urethane and styrene foams are impermeable to water vapor by virtue of their closed-cell structure. U-F foam does not form its own vapor seal, however.

The colder the outside temperature is, the more important a good vapor seal becomes. Since all vapor barriers pass some water, double barriers are often used when the winter temperature gets below minus-10 degrees F. The second barrier—usually plastic film—stops most of the vapor that gets past the first. This prevents the condensation that could result on the super-cold sheathing surface.

DETERMINING YOUR INSULATION NEEDS

Most houses built before 1945 don't have any insulation. After 1945, insulation slowly started to take hold, showing up first in attics. Finally, around 1955, builders started putting insulation in walls too. But often they used too little insulation. How can you check your home for insulation? Some places, such as your attic, are easy to check. Just look. If there's insulation there, measure its thickness and compare it to recommended minimums. If it's below the minimum you should add to it. When you are adding to the cold side of existing insulation, use insulation without a vapor barrier; otherwise moisture may condense between the layers. Unfaced blankets or batts, or loose fiber fill are good choices.

Other spots, such as crawl spaces, basements, ceilings over unheated porches or garages are usually easy to check by eye. But when you get to finished walls, or finished ceilings, you just can't see in there to check for insulation.

For a quick check try this: Place your hand against the inside of a suspected exterior wall. Note the temperature. Then place your hand against an inside partition wall—a wall that doesn't separate the inside of your home from the outside. If the exterior wall is insulated, it should feel close to the same temperature as the interior partition. If it feels appreciably colder, suspect that it isn't insulated. For a further check, remove an electrical outlet from the exterior wall. If there's a gap around the metal junction box behind the outlet, shine a flashlight through the gap and look for insulation. If there's not enough of a gap to see through, you can slice away the wallboard a little with a knife. Just be sure you don't cut away so much wallboard that the outlet cover plate can't hide the resulting opening.

If you find any insulation at all in your walls, it probably won't pay to add to it—the job is too expensive. But if there is no insulation, the picture changes. And any time you find insufficient insulation in a spot where you can add to it yourself, go ahead and do the work. Your labor is free so labor won't contribute to the final cost of the installation.

HOW TO INSULATE YOUR ATTIC

Unfinished Attic, No Floor. This is your simplest insulating job, and also the most important. It may save you more

than $200 each year in heating and cooling costs. Best insulation for the job is mineral wool batts or blankets with a paper vapor barrier. These will install easily between the floor joists or trusses in your attic. Check the map: "Which R-values Go Where?" shown earlier in this chapter. Note: Insulating between the floor joists is the best way to handle an unfinished attic—if you never plan to finish that attic. If you plan to finish the attic in a year or two, you'll be better off insulating between the rafters, as described later. If you have no definite plans for the attic you can still insulate between the joists. Since the insulation is simply placed in position —not stapled—you can always remove it later and put it between the rafters. Insulating between the joists requires less insulation because there's less area to cover. And it saves more on heating and cooling because

When installing insulation in an attic floor, run it to the edge of the top plate. Fill the gap with loose insulation and add a baffle to prevent the insulation from blocking air flow from soffit vents. Ventilation is important in removing heat and moisture.

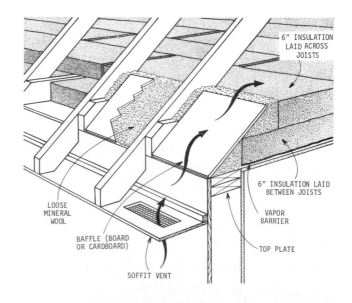

it doesn't require that you heat or cool the attic.

Start the job by placing a few planks across the floor joists to provide safe footing. Don't step between joists or you'll go right through the ceiling below. String up some form of temporary lighting and measure the area of the attic floor. Buy enough insulation to cover this area. Usually you can get by with slightly less than the full area measured, because some of this area is filled by joists. Figure your insulation need as 90 percent of the floor area if joists are 16 inches on center, or 94 percent for 24-inch spacing.

Before placing the insulation, seal all places where wiring, pipes, or plumbing vents penetrate the attic floor. Duct tape or caulking rope will do the job. Check the underside of the roof for any waterstains that indicate roof leakage. You should fix any leaks before insulating; wet insulation is useless. Now place the batts or blankets between the joists, paper side down. If your home has eave vents, be sure not to block the flow of air from the vents into the attic. (See the drawing on page 14.) If there are any light fixtures protruding up through the ceiling, keep insulation at least three inches away from them. This will allow them to cool themselves when they're turned on. Insulation would trap the heat they produce and possibly cause damage, or even a fire. If a chimney passes through your attic, the space between the chimney and any framing around it should be filled with insulation, but only noncombustible insulation. If your batts have a paper vapor barrier, remove it from the wool you stuff into this space,

Slit the ends of the batts where they butt against joist bridging (See

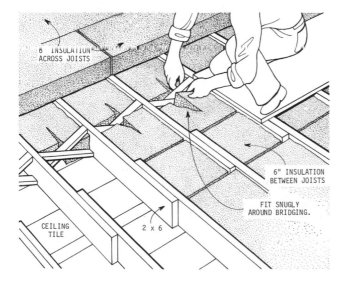

Cut ends of blankets to fit snugly around any bridging between the floor joists.

6" INSULATION
ACROSS JOISTS

6" INSULATION
BETWEEN JOISTS

FIT SNUGLY
AROUND BRIDGING.

CEILING
TILE

2 x 6

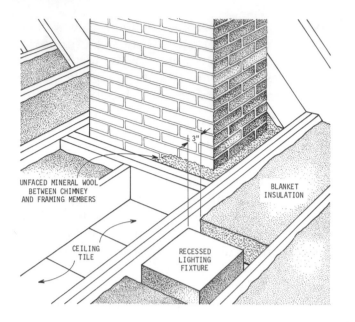

UNFACED MINERAL WOOL
BETWEEN CHIMNEY
AND FRAMING MEMBERS

3"

BLANKET
INSULATION

CEILING
TILE

RECESSED
LIGHTING
FIXTURE

Stuff the space between a chimney and surrounding framing members with un-faced mineral wool. If a recessed light-ing fixture protrudes up through the ceiling, keep insulation three inches away so the fixture can dissipate heat.

page 15.) It's not a bad idea to pour a little loose material wool insulation over spots like this where the batts are broken or where they fit poorly. If you are insulating to a value higher than R-22, you'll need two layers of insulation. Use batts or blankets without a vapor barrier for the second layer, and place it crosswise to the first layer.

Unfinished Attic with Floor. This job is a little trickier than insulating an attic with no floor. You have no access to the space between your floor joists, so you have to put the insulation else-where. You could have the underfloor space insulated by a contractor, but this will cost about double what you'll have to pay to do the job your-self. The contractor will blow insula-tion in under the floor, and since this is a blind area, you'll never really know how completely the underfloor space has been filled. You're better off doing the job of insulating your-self, putting batts or blankets be-tween the roof rafters, across the col-lar beams, and against the attic end walls.

Collar beams are 2x4s that run be-tween the rafters at a height of about eight feet. If there are none in your attic, you should install them. Just buy 2x4s, cut them to the proper length, and spike them to the sides of the rafters. Be careful to make these beams all the same height from the floor, and be sure they're level. If you ever install a ceiling in the attic it will fasten to these beams. The open space above the collar beams pro-vides for ventilation, helping keep your attic dry, and cooling it in the summer.

Buying Insulation.   Use foil-faced mineral wool 6½ inches thick for the space between the collar beams. This is minimum. For more, use unfaced batts over these batts to get a higher R-value. Or use an extra-thick batt. For the end walls, use foil-faced blankets thick enough to fill the space between the studs. This will almost always be 3½-inch or R-11 blankets, minimum.

Between the rafters use insulation that will leave a ventilating air space between insulation and roof. If your rafters are mere 2x4s, you'll have only enough space to fit R-13 mineral wool. To provide more R-value than that, you can fasten foam insulation to the faces of the rafters. But be sure to cover the foam with a fire barrier such as ½-inch fire-guard gypsum board. To avoid condensation, use mineral wool without a vapor barrier.

With 2x6 rafters, install the batts between the collar beams first. Use a staple gun and fasten the flanges of the foil vapor barrier to the edges of the collar beams. The vapor barrier should face toward the inside of your home, toward the heated part. Staple the batts between the rafters in the same way, and do the same with the blankets between the studs on the end walls. At all times, be careful to cut the insulation so it fits snugly. Pay special attention to the ends of the blankets and batts.

If you're adding insulation to existing but insufficient insulation over the collar beam, do not use insulation with a vapor barrier. If you can't ob-

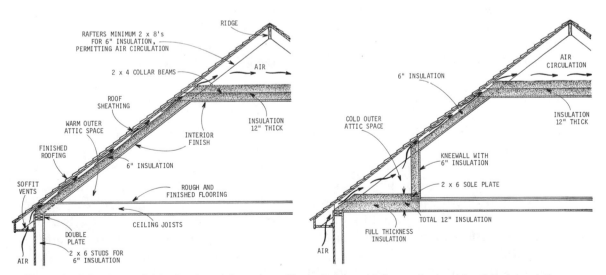

Two ways to insulate a finished attic are shown here. The technique, at left, uses the least insulation. Insulating kneewalls, at right, requires more insulation. But less area is heated; so savings will be larger over the long haul.

tain blankets or batts without a barrier, slit the new barrier every foot or so with a knife so that it won't condense vapor between the old and the new layers.

If you are adding to insufficient insulation between the rafters, cut the stapling flanges on the old blankets and push them back into the spaces between the rafters. Slit the vapor barrier if any, then staple the new insulation—with barrier—over the old.

Finished Attic. This is a lot like insulating an attic with a floor, except that access to the spaces to be insulated is a problem. In many cases you'll have to cut a hole in the ceiling, and into each of your sidewalls, but the holes won't be too hard to repair, and the resulting savings will be worth the effort. Your only other choice is contractor-installed blown-in insulation. As mentioned before, this will cost you about twice your do-it-yourself expenses.

Look at the preceding drawing of the finished attic. The best way to insulate the finished attic is with batts in the ceiling, and between the rafters from the ceiling down to the floor. This uses less insulation than your alternative: insulation in ceiling, between the rafters down to the knee walls, then down the knee walls and across the outer floor. And it insulates the outer attic spaces so you can use them for storage if you like.

Your first step is to gain access to the spaces that need insulation. If there is no other way up there, cut a hole through the ceiling, large enough to climb through, and get up into the space above the ceiling. If there is no way to get into the outer attic spaces, you'll have to cut through the walls to reach these areas, too. Once you have access the hard part is past. You and a helper can now insulate over the ceiling and in between the rafters without much trouble. The two of you should climb up through the ceiling with a supply of foil-faced batts. Have your helper crawl out into one of the outer attic spaces. Now you can slip the batts down through the spaces between the rafters. Sometimes shingle nails may extend all the way through the roof sheathing and snag the batts as you try to slide them down between the slanted ceiling and the roof. If so, your helper can aid you by reaching up and pulling the batts while you push.

If he can't reach the batts, try this trick: Take two strips of wood as long as the width of the batts. Using a small clamp, fasten these to the end of the batt, making a sort of insulation sandwich. Now fasten a length of wire to the clamp. You can then pass the wire down to your helper and he can pull the batt into position. Remove the wood strips and fasten the insulation in place, stapling the foil flanges to the edges of the rafters. When all the rafter spaces from ceiling to floor are filled, insulate between the ceiling framing members, and climb back down through the hole.

You can repair the hole to match the ceiling, or simply make a lift-out trap door, trimming the edges of the opening with ordinary clamshell casing, available at any lumber store. The openings in the walls can be handled the same way. Small doors here are a good idea since they give access to the outer attic spaces for storage.

Note, if you want or require extra insulation over the ceiling, install a second layer of mineral wool over the first, and at right angles to it. Use no vapor barrier on this layer. If you can't get insulation without a barrier, slit the barrier to allow vapor to pass through.

Access to the attic end walls will be a problem. Best advice here is to forget about them for now. If you hire a contractor to blow insulation into the frame walls of your house, he can get the attic end walls at the same time. If you want to do the job yourself you have a few options. You can remove the finished wall surface from the studs and insulate between the studs with blankets. Then replace the wall. If you are careful removing the wall surface you can often reuse it, especially if it is paneling. You can hide chips along the seams with strips of color-matched molding. Taping the seams and nail holes in drywall will take care of damaged sheetrock if the damage isn't too severe.

Another choice is to cement styene foam over the present end walls; then cover this with at least $1/2$-inch gypsum board. I don't like this solution myself, because it adds $1\frac{1}{2}$-inches of bulk to the walls, and this causes trim problems around windows, doors, outlets, and so on.

A third choice is to cut a two-inch hole in the wall between each stud, about a foot below the ceiling so that you have enough room for pouring. Then you can fiddle around pouring vermiculite or perlite in through these holes until it fills the stud space up to the level of the holes. Then patch the holes. Of course this leaves the top foot of the wall uninsulated, but it's better than nothing at all.

One last option. Cut away about a foot of the wall surface all across the end wall, just below the ceiling. Then cut away about two inches of wall all across the wall at floor level. Then using the trick for pulling insulation into place with a wire described above, you can pull R-11 batts or blankets down through the space at the top of the wall until they reach the floor. Staple the insulation at the top to prevent its falling down, repair the missing section at the top of the wall, cover the gap at the bottom of the floor with a base molding.

HOW TO INSULATE YOUR CRAWL SPACE OR BASEMENT WALLS

Crawl Space. Here you insulate only the walls. This is a quick and easy job where appearance of your work doesn't matter. If your crawl space has a dirt floor, you'll want to cover it with a plastic vapor barrier, 4 to 6 mils thick. The only condensation that collects later is *under* the barrier, and this then soaks back into the ground. Lay your vapor barrier after you place the insulation so you can avoid trampling the plastic and possibly tearing holes in it. If your crawl space has a concrete floor, theoretically, the builder should have placed a vapor barrier under the concrete, so you can skip the plastic film.

Two of the walls in your crawl-space will be perpendicular to the floor joists; the other two will run parallel to the joists. You should use different techniques to insulate the two different types of walls. When the walls are perpendicular to the joists, cut short sections of insulation (blankets without vapor barrier) to fit snugly between the sill, subfloor and joists as shown. Then install the strips of insulation covering the wall and perimeter of the floor. Secure these with nails driven through lath or strips of wood about $3/4 \times 1 1/2$ inches. The blankets should butt up tight against one another, and extend about two feet in towards the center of the crawl space.

Where the walls run parallel to the joists, just install the insulation strips in one piece as shown in the drawing on page 21. Use the nailing strips as you did on the other two walls to hold the insulation in place.

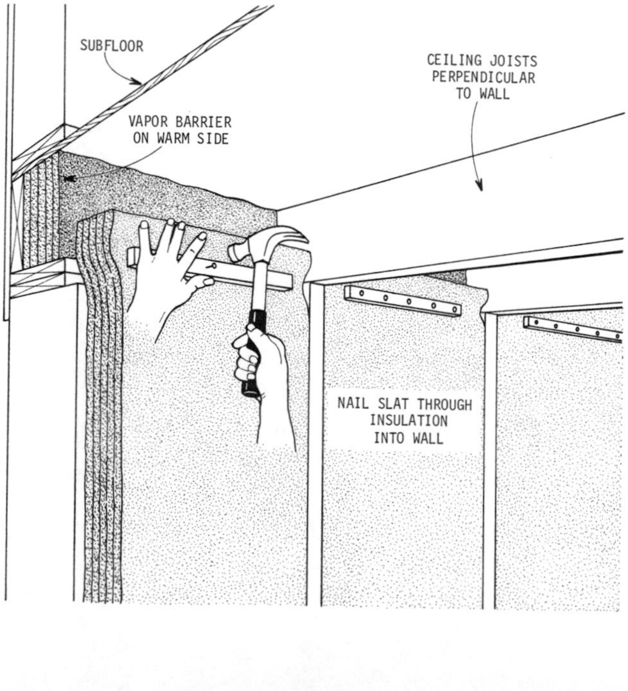

Where floor joists above a crawl space run perpendicular to the wall you are insulating, install the insulation as shown here. Cut a piece of blanket to fit snugly against the header. Then fasten strips of insulation to the sill and run them down the wall and two feet out over the crawl-space floor.

SUBFLOOR

CEILING JOISTS PERPENDICULAR TO WALL

VAPOR BARRIER ON WARM SIDE

NAIL SLAT THROUGH INSULATION INTO WALL

Once the insulation is in place, you can add the vapor barrier if required. Polyethylene film laid out in strips across the ground will do the job. Sometimes you'll see advice recommending that you tape the seams where one strip of the film overlaps the next, but polyethylene is waxy and nothing sticks to it very well. You're as well off to overlap the strips about six inches and let it go, or maybe weight the overlap with wood scraps or bricks. Be sure to tuck the film under the batts and up against the wall. Then use 2x4s or rocks to weight the blankets next to the wall and hold them in place. You won't be able to avoid stepping on the vapor barrier as you finish up this job, but at

least try to keep your trampling to a minimum.

Basement Walls. This isn't your easiest insulating job. The problem is hiding the insulation once it's in place. To do this you have a couple of choices.

Foam boards. A technique that avoids a lot of work employs plastic foam boards. Inch-thick boards will do the job. You can cement them directly to the concrete walls of your basement, then cement ½-inch gypsum board over the foam to provide a finished wall, and fire protection.

Be sure to use the right cement for this job. Many mastics will melt the plastic. Check the label on the adhesive to be sure it's compatible with foam. Foam boards usually come 2 by

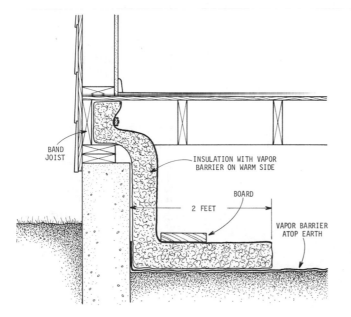

BAND
JOIST

INSULATION WITH VAPOR
BARRIER ON WARM SIDE

BOARD

2 FEET

VAPOR BARRIER
ATOP EARTH

When floor joists run parallel to the wall being insulated, install the insulation as shown here. Place a board on top of the insulation near the wall to keep the blankets snug against the wall.

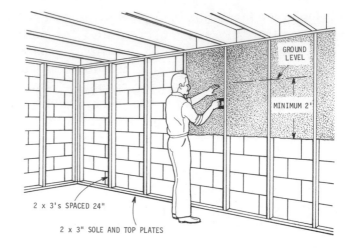

GROUND LEVEL

MINIMUM 2'

2 x 3's SPACED 24"

2 x 3" SOLE AND TOP PLATES

Basement walls can be insulated as shown here. Build a frame wall of 2x3s spaced 23 inches on center. Place blankets between studs and run them at least two feet below outside ground level (all the way to the floor in cold climates). Then panel or sheetrock over the 2x3s.

4 feet. Just cement these panels to your walls, cutting them whenever necessary with a saw, electric carving knife, or an old kitchen knife heated with a torch. The hot knife eliminates a lot of sawdust that will cling to everything in the area due to static electricity. Cementing the gypsum board over the insulation goes quickly, then taping and finishing off the joints will be the most time-consuming part of the job. Still, this technique is a little easier than the next one. (Caution: In extreme northern areas such as Minnesota, Alaska, and northern Maine, this method may cause frost heaving in the foundation.)

Stud walls. You can also build a stud wall of 2x3 lumber against your existing walls, to install insulation blankets between the studs, and then nail on a finish wall. The advantage to this system is that you can use paneling instead of drywall. (Paneling is not recommended over plastic foam because it is not fire resistant.) Studs and blanket insulation give you a wider choice of wall finishes. And paneling is a little easier to install than drywall because you can skip all that tedious joint taping. The drawback is the expense and bother of adding the 2x3-stud wall.

To start the wall, fasten a 2x3 bottom plate to the basement floor, all around the perimeter of the basement, tight against the wall. Use masonry nails about three inches long. Then nail a top plate to the bottoms of the joists overhead, also tight against the wall. Both plates should be placed flat, not on edge. Then toe nail in vertical studs spaced 24 inches on center, and cut to fit snugly between the two plates. Space these studs

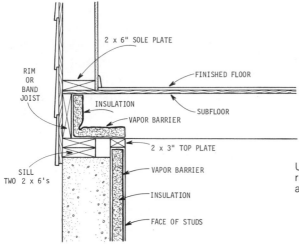

2 x 6" SOLE PLATE

RIM
OR
BAND
JOIST

FINISHED FLOOR

INSULATION

SUBFLOOR

VAPOR BARRIER

2 x 3" TOP PLATE

SILL
TWO 2 x 6's

VAPOR BARRIER

INSULATION

FACE OF STUDS

Use small pieces of insulation to cover the sill and rim, or band, joist. Just cut the pieces slightly oversize and wedge them in place.

carefully, so the paneling or gypsum board you'll nail on later will fit properly.

Note: Needless to say, if your basement walls leak, you should take care of this problem before you begin the insulating job. If the walls weep over a wide area, a couple coats of cementitious paint such as Thoroseal should solve the problem. If not, try one of the new epoxy coatings such as Zap or Epoxite. Leaking cracks can be repaired with hydraulic cement, even if they are running water at the time you make the repair.

Once the stud wall is in place, staple insulation in between the studs. Use foil-faced blankets, with the foil facing in. In moderate climates you can get by running the insulation to just a couple feet below the level of the ground outside. Northern climates make it worthwhile to run insu-

lation all the way to the floor. Caution: In extreme northern areas (Minnesota, Alaska, and northern Maine) this method of insulation should not be used. As with the foam board method described above, this method may cause frost heaving in the foundation. If you live in extreme northern country, consult your local HUD/FHA field office for advice.

To complete the insulation job, tuck short pieces of insulation up over the sill and the top plate you've added, to insulate the sill and the joist header or band joist. See the accompanying drawing for details.

Last, apply drywall or paneling over the joists to finish off the wall. Paneling will be the easiest to install, but it will cost at least twice as much as gypsum board. Sometimes you can find 4x8 sheets of paneling for as low as four dollars; gypsum board runs around $2.40 and up. If you're not too concerned with the look of the basement walls you can use gypsum board and leave the joints untaped until you feel like tackling them.

HOW TO INSULATE UNDER A FLOOR

Any home or part of a home that sits over an open, unheated area will benefit from insulation in the floor. Post and beam houses, homes built on piers, or homes with rooms over unheated porches or garages all fit in this category. Insulating under a floor is an easy job, but it does present one small problem. Since the insulation should be installed with the foil vapor barrier facing up, the stapling flanges will be on the top of the insulation, out of reach and useless for fas-

tening the batts in place. So you'll have to use chicken wire stapled to the bottom edges of the floor joists to hold the insulation against the floor.

Currently, some authorities recommend only R-11 blankets or batts for floor insulation. But there's room be-

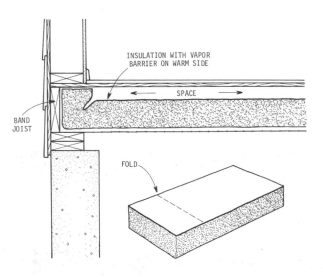

Insulation in floors over unheated spaces goes in with the vapor barrier facing up, so stapling it in place is impossible. Use chicken wire stapled to the bottoms of the joists to hold the blankets in place. Fold the ends of the blankets, as shown above, to insulate the rim or band joists.

tween the joists for thicker batts, and with heating costs rising, R-22 or R-19 insulation will certainly pay for itself in northern climates.

You'll be able to do a neater, more effective job if you can round up a helper. Buy chicken wire or any kind of wire mesh equal in width to your joist spacing (24 or 16 inches). Start at a wall, stapling up a couple of feet of wire under a pair of joists. Have your helper slide the end of a roll or batt up over the wire (foil side up). Fold the end of the insulation, as shown. This will insulate the joist header. Now as your helper unrolls the blanket of insulation and shoves it up between the joists, you follow along behind him stapling the wire in place.

If you can't get help, run the wire across the joists, from one edge of the floor to the other. After you staple a couple rows in place, slide short sections of insulation in on top of the wire. Add a couple more rows of wire; then slide in more insulation. This technique is more work than the first, and it results in frequent gaps in the moisture vapor barrier. This isn't too critical, though, because the main purpose of the foil barrier in this particular application is simply to prevent radiation heat loss.

INSULATING FINISHED FRAME WALLS

This is definitely a job for an insulation contractor. Getting insulation into finished walls involves blowing or shooting the stuff, and you just don't have the equipment required.

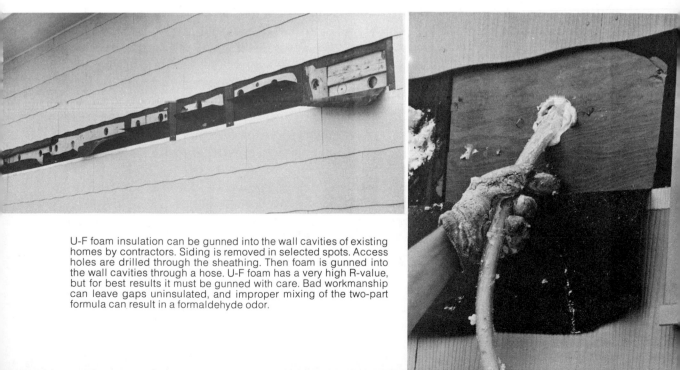

U-F foam insulation can be gunned into the wall cavities of existing homes by contractors. Siding is removed in selected spots. Access holes are drilled through the sheathing. Then foam is gunned into the wall cavities through a hose. U-F foam has a very high R-value, but for best results it must be gunned with care. Bad workmanship can leave gaps uninsulated, and improper mixing of the two-part formula can result in a formaldehyde odor.

You have a few choices here. Blown-in insulation types include mineral wool and cellulosic fiber. The mineral wool will give an R-value of about 8 in a standard 2x4 stud wall. Cellulosic fiber will rate about R-10 to R-13. Gunned-in ureaformaldehyde foam, probably your best choice, will give you a rating around R-15 to R-20.

If I were having my walls insulated, I'd definitely go with U-F foam. No matter which material you choose, much of your costs will be for labor. So you might as well get as much R for your money as possible.

Saving Money. A great deal of the labor costs for installing any blown or shot-in insulation are for preparation and cleanup. Sections of your home's siding are removed, and holes are drilled through the sheathing to give access to the stud spaces to be insulated. Often, you can cut your costs by doing this work yourself, then replacing the siding when the job is done. Check with the contractor you choose to see if he'll go along with a deal of this type.

Contractors. Picking a good one is important. Application of any of these forms of insulation takes skill and know-how. An inexperienced or shoddy worker can leave large uninsulated gaps, and these are hard to detect. So check any contractor out before you hire. Start by getting estimates from three or four who have

been recommended by friends, building associations, your banker, or nonprofit home improvement assistance centers. Check each of them out with the Better Business Bureau. Ask them for lists of past customers and check with them for their opinion. When you've chosen a contractor, have him put the contract in writing. Insulating the walls of the average home can cost around $1,000. When you're spending that kind of money it pays to be careful.

Moisture-vapor Barrier. None of these blow-in or shoot-in insulation materials provides any moisture vapor barrier, not even the U-F foam. So you'll probably have to provide a moisture barrier of your own if you have your walls insulated. The way to do it is unconventional compared to the ways discussed so far. First step is to paint the inside of all insulated walls with a low-permeability paint. Glossy paints will do the job but most water-thinned finishes breathe to some extent, so oil paints are the best choice. It's not likely you'll want a glossy finish on your walls, however. So after you put on the enamel you can overcoat it with an ordinary flat wall paint. Make sure the oil-base paint is close to the same color as your finish coat or you may have to apply two topcoats to get complete coverage. In addition to the paint treatment, caulk openings in the inside wall that could allow moisture to get inside the walls. Window and door frames are especially critical. These areas are abnormally cold and provide perfect conditions for condensation.

WHY HEATED, UNINSULATED ROOMS FEEL CHILLY

One odd characteristic of an uninsulated house is that it can be heated to a high temperature, yet you still feel cold. That's because the exterior walls are cold. Your body radiates heat and the cold walls retain the heat. It's as though the walls were sucking the heat right out of your body. If you try to compensate by raising the thermostat, you get little relief. You can still lose heat to the walls, yet the air inside the house feels stuffy.

Insulate, and you may feel comfortable at 68 degrees in shirtsleeves. In an uninsulated home you can feel a chill with the thermostat cranked up to 75. Of course you can simply stay away from the exterior walls in your home, and you'll feel warm despite their lack of insulation. But you lose a lot of living space if you do this. Clearly it pays to insulate.

CAULKING AND WEATHERSTRIPPING

Insulation slows down heat loss through walls, roofs, floors. But it won't stop cold air from leaking in through cracks around doors, windows, airconditioners, and exterior trim. The job of stopping cold air infiltration falls to weatherstripping and caulk. Caulking and weatherstripping your home is easy. Even if your home has no caulking or weatherstripping, you shouldn't have to pay more than $80 to bring things up to snuff. And you'll get your money back in savings on heating and cooling in less than 12 months in most cases.

There are other reasons to caulk and weatherstrip. Caulking will prevent the seepage of water into your home where it can cause rot and paint peeling, and where it can soak insulation, rendering it ineffective. Both caulking and weatherstripping will prevent drafts that can chill your body even when indoor temperatures are above 70 degrees. And both can keep out the dust and dirt that always manage to sift into your home through every available crack and crevice.

DO YOU NEED TO CAULK AND WEATHERSTRIP?

Maybe not, if your home is relatively new, or if it has been painted recently. Most modern windows are well weatherstripped at the factory, and they shouldn't need weatherstripping replacements for years, if at all. And a good painter always caulks a home when he paints it. So before you go out and stock up on supplies, give your home a check to see if it really needs to be sealed. Weatherstripping is easy enough to spot. Just look for it around door and window openings. It may take any of several forms, but it will be clearly visible if you open the suspected door or window. If the stuff is there and in good shape, good. If not, you have a job on your hands.

To check for caulk, just look over the seams in the exterior skin of your home. Check around door and window trim, between chimney and house, etc. There should be no visible cracks that can allow the passage of air. Caulk should be in good shape, with no breaks. If not, caulk.

Buying Caulk. There was a time when the only caulk you could buy was oil-based stuff with a life expectancy of about two years at the most. Today there are much better materials on the market. Oil-base caulk is still available, and it's still the best seller due to its low cost. But it's now foolish to use it. It will dry out, shrink, crack, and fall out of place in a couple years. Then you'll have to scrape it out and replace it. Despite the fact that it sells for under a buck a tube or cartridge, it's still a bad buy.

First choice. Buying caulk is really very simple. Best advice is to get urethane caulk. The stuff you want is Vulkem 116. It comes in standard 11-ounce cartridges that drop into ordinary caulking guns. It guns with ease, and gives you a nice neat bead. It sticks to just about anything, and it lasts for years. It's nontoxic and you can paint over it. It will fill almost any size crack, and it works on moving or static joints. On top of all this, it costs only about $2-3 a cartridge, about the middle of the road in caulking cost. In other words, this one caulk does it all. You don't have to worry about whether a crack is too wide, or if a joint is a moving joint, or whether the caulk will stick to that substrate.

You do have one worry, though. This stuff isn't so easy to find. You may have to go to the Yellow Pages under Calk (not caulk) and look for a Vulkem distributor. If that's a dead end you can write to Vulkem Sealants Div., MAMECO Int., 4475 E. 175th St., Cleveland, OH 44128. It's worth the effort.

Second choice. Silicone rubber is your second choice. It does a good job, but it is a little harder to apply than urethane. It costs more, requires a primer on porous surfaces, and in many cases won't take paint. If you do buy this stuff, get General Electric's. This is one silicone you can paint over.

Other choices. If for some reason you can't get urethane or silicone caulk there are a few other choices. Hypalon is a good third choice. And a premium acrylic latex is a good fourth choice. But make sure it's a premium grade. Macco Super Caulk, for example, has a 20-year guarantee. Many of the cheaper goos have no guarantee at all. And even the best acrylics won't cover the complete range of applications that urethane, silicone, or Hypalon can.

Where to Caulk. Quite simply, you should caulk any seam on the outside of your home that can let air leak in or

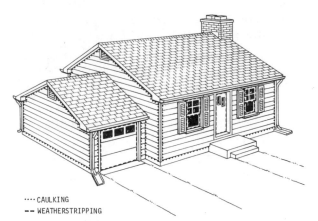

···· CAULKING
── WEATHERSTRIPPING

Careful caulking and weatherstripping can reduce heat losses as much as 25 percent. Use caulking at stationary joints; use weatherstripping at moving joints such as those around doors and windows. Caulking of joints at the ends of unheated attics has no weatherproofing value, although it will help to keep insects out.

out. To be more specific, here's a list of common caulking spots:

- Around door and window frames: top, bottom and sides.
- Where siding meets the trim at house corners.
- All around the sill (where wood meets foundation).
- Between chimney and siding.
- Between porches and house.
- At sillcocks, around dryer vents,

outside electrical outlets—wherever there are breaks in the siding.
- Between gable and roof eaves if attic is heated.
- Where storm windows meet window frame.
- Inside the home, wherever pipes, wires or vents penetrate a ceiling below an unheated attic.
- In any cracks or unputtied nail holes or other openings you might find in the exterior of your home.

How to Caulk. First you'll need a caulking gun and a supply of caulk, preferably urethane. How much caulk will you need? This varies with the size and number of voids to be filled. You can figure roughly on using one cartridge for every two doors or windows, four cartridges for the sill, one cartridge at each corner of the house, and a couple for a chimney. These are rough estimates. It won't hurt to overbuy if you can return unused cartridges, but urethane caulk has a shelf life of about two years. After that it starts to set up in the cartridge. So don't count on keeping the extra caulk around for more than a couple years.

If you have window airconditioners, you should get some caulking cord to seal around them. This sticky cord can be pressed in place to make a weathertight seal whenever you want it, yet it peels off when you remove the airconditioner for the winter. You can also use it to seal windows that you won't open during

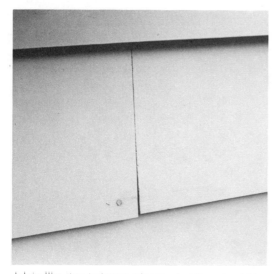

Joints like this between pieces of siding should be caulked, not only to reduce air infiltration, but to prevent seepage of moisture which will then cause paint and siding failure.

Large gaps like this one between the bottom strip of siding and the concrete foundation wall should be filled with a material such as oakum, then caulked. A wide crack may require a double bead of caulk.

the winter. Opening the windows, of course, will break the seal.

Before you buy your caulk, check the seam at the sill of your home. If it's more than about ⅜-inch wide it will pay to stuff it with oakum or strips of mineral wool insulation, before you caulk it. A good hardware store or plumbing supplier will carry oakum rope. You may have enough insulation scraps around the house to do the job.

Preparing the surface. A good caulk such as urethane will stick to just about anything, but it does need a solid substrate. You can't expect it to adhere over chipping or flaking paint, or over old oil-base caulk that's dried out and falling away. So start your caulking by cleaning out any built-up paint, old failing caulk, dirt, etc. A putty knife, heavy screwdriver or an old chisel will do the job. If your house is quite dirty, take a trip around the house with a stiff brush and a bucket of detergent a day or two before you start caulking. Rinse your work as you go with a garden hose.

Doing the Caulking. When you're ready to go, load a cartridge in your caulking gun, snip or slice the tip off the nozzle at a slight angle, and puncture the seal inside the cartridge by jamming a piece of wire down the nozzle. Squeeze the trigger on the gun a couple times, and the caulk should start into the nozzle. Before it starts to

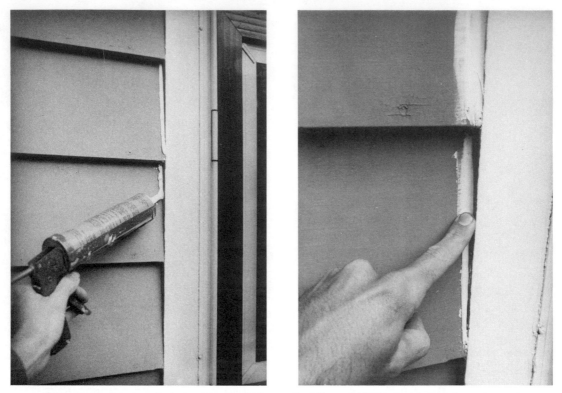

Seams around door and window trim should be caulked. Run a fairly thin bead of caulk as shown at left. Then go over the bead with your finger, shaping it into a neat cove as shown at right. After caulk has dried you can paint over it.

ooze out, place the nozzle in position, squeeze again and begin laying a bead. The trick to an efficient, neat job is to balance the rate of flow against the speed with which you draw the nozzle along a seam. You want to be sure to make a wide enough bead to cover both sides of the joint, but you don't want caulk all over the place. Too much caulk is a waste, and it gets messy. If you do run too big a bead, you can go over it with your finger, smoothing it out and squeegeeing off the excess. A little water on your finger will help make urethane caulk even smoother. You can remove the caulk from your hands with a cloth if you act before the stuff starts to set up.

Once you start to run a bead, it pays to keep going on that run without a break. Once you stop a bead, it's hard to start it again smoothly. Sometimes the caulk will start to run away from you—running out of the nozzle faster than you can place it. If it does, twist the plunger on the gun. The ratchet drive mechanism will disengage and take the pressure off the caulk which will then stop flowing. Then you can reengage the ratchet, squeeze the trigger a few times and continue with your work.

Needless to say, at some point in your caulking, you'll have to climb a ladder. Don't try to stretch and strain to reach your work. Move your ladder whenever your work starts to get out of reach. Stretching to reach your work will result in a sloppy job, and it's dangerous, too.

HOW TO WEATHERSTRIP YOUR HOME

When you weatherstrip, you are concerned with two basic targets: windows and doors. Materials and techniques are different in each case, so we'll treat them separately. Let's start with windows, since they're usually the easiest to handle.

Window Types. The weatherstripping you choose for your windows will in large part be determined by the type of windows you have. Conventional *double-hung* windows—the type that

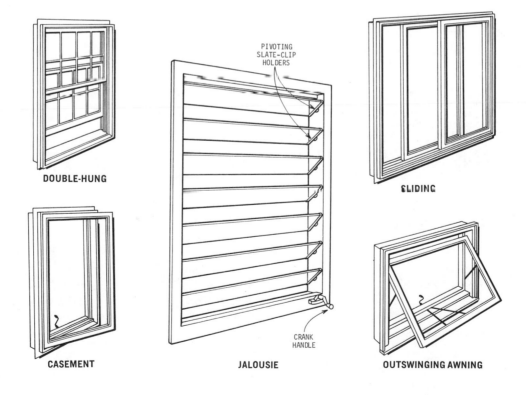

These are the five most common types of adjustable windows. Tips on weatherstripping are given in the text.

slide up and down—are best handled with rolled vinyl weatherstripping. This stuff is easy to install, and it lasts a long time. About the only drawback is that it shows after installation. If your windows need weatherstripping (most modern windows are weather-stripped at the factory), rolled vinyl is a good choice. The drawing on page 35 shows how to install it.

Another durable material is thin spring metal. In many cases this is only slightly harder to install than the rolled vinyl, but if your windows are already a tight fit in their channels, there may not be enough space for the strip. If so, you're better off with rolled vinyl. The chief advantage of spring metal is its ability to withstand the friction caused by opening and closing a window. If you tend to leave your windows closed most of the time, sliding friction is little problem and rolled vinyl gets the nod. On the other hand, metal is a better choice for windows that get a lot of exercise. It may be worth your while to use it, even if you have to remove the win-

Most modern windows come weatherstripped from the factory. Here are metal channels in a readymade bay window. Channels like these are available for converting unstripped sashcord style windows of the past. You install them by removing window and sash-cord, setting channels in place, replacing the window.

Metal-and-felt weatherstripping is unsightly and tends to wear out quickly, as this strip testifies. It should be replaced, preferably with a more durable and effec-tive type of stripping such as a combination vinyl tube and aluminum strip.

dow sash from its channel, and plane a little off the edges to make room for the weatherstripping. Installation is then a matter of tacking the metal in place as shown.

Foam vinyl with an adhesive backing is another material you can use on double hung windows, but only for sealing at the top and bottom. It will wear out almost instantly if exposed to the sliding edges of a window.

If you have double-hung windows that are old, loose in their channels so they rattle in the wind, unweatherstripped, and otherwise on the brink of becoming obsolete, you might consider installing new window channels. These will modernize the win-

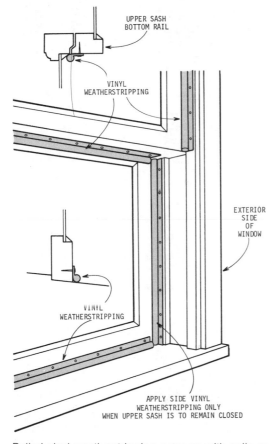

Rolled vinyl weatherstripping goes up with nails or thumbtacks supplied with the stripping. Apply it to double-hung windows as shown here. Note placement of the strip on the bottom of the upper sash.

Here Quaker City window channels are held secure over old window sashes while being installed in the old frame. The channels are then simply nailed or screwed in. The old parting bead is reinstalled.

dow (eliminating sash cords or chains) and weatherstrip them, solve the rattling problem, and bring your windows up to date without the expense of buying all new windows.

Installation involves prying off the parting bead and removing your old sash. You also take out any old weatherstripping, spring metal, and sash cords. You slip the new metal window channels around the old sash and replace it in the window opening. Then you nail the new channels in place and replace the old parting bead. One such window improver is the CBW Side-Ride Window Channel by Quaker City Manufacturing, Sharon Hill, Pennsylvania.

Casement windows and *awning windows*—those that open like doors —are a lot easier to weatherstrip than double-hung windows because their hinge mounting causes no friction to wear out weatherstripping. Adhesive-backed foam applied to the outer surfaces of the stops the window butts against when you close it will serve well. The foam stripping is cheap, and you can do a whole window with it in a minute or two.

Sliding windows—essentially double-hung windows turned on edge— should be treated like double-hung types, as described above.

Jalousie, one last type of window, requires a special weatherstripping that's a clear vinyl channel that slips over the edge of each slat of glass. A similar channel type stripping is used on some casement windows, but most of the better casements come weatherstripped when you buy them.

Plastic foam sheathed in vinyl is easy to install, but tacks have a tendency to pull out. Stripping should be pressed lightly against the closed door when you seat the tacks. Too much pressure will result in the weatherstripping being pinched between door and jamb. Too little pressure gives a poor seal.

WEATHERSTRIPPING DOORS

This is a job that would present few problems if it weren't for the friction at the lower edge of the door. Almost any material will take care of the sides and top of the door. Adhesive-backed foam, rolled vinyl (with or without an aluminum backing strip), spring metal, or foam rubber on a wood backing strip are all easy-to-apply alternatives. Of these, the spring metal is the most durable. It should

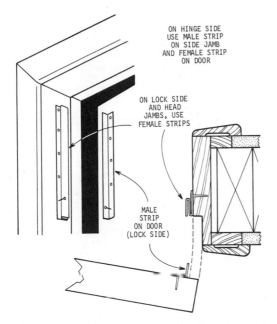

ON HINGE SIDE
USE MALE STRIP
ON SIDE JAMB
AND FEMALE STRIP
ON DOOR

ON LOCK SIDE
AND HEAD
JAMBS, USE
FEMALE STRIPS

MALE
STRIP
ON DOOR
(LOCK SIDE)

Interlocking metal channels must be aligned properly for proper mating of male and female strips. Do the head of the door first, nailing the male strip to the door and the female strip to the door frame. On the hinged side of the door (not shown), mount the female strip on the door, the male strip on the jamb. Do the lock side of the door last, male strip on the door, female strip on the jamb.

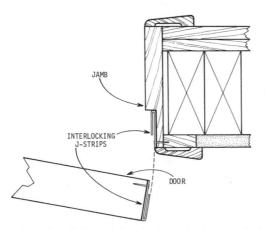

JAMB

INTERLOCKING
J-STRIPS

DOOR

Proper installation of interlocking J-strips requires routing a rabbet in the edge of the door and jamb as shown here. This makes installation difficult but protects the stripping from damage.

last a lifetime unless it gets bent and snags on the door.

If you're skilled with tools, a couple other forms of metal weatherstripping may appeal to you. These are interlocking metal channels and interlocking J-strips. Both provide an excellent seal, and both are very durable. The *metal channel-type* is exposed during installation, and can possibly be damaged then, but it's the easier of the two to mount. Proper alignment is the only real problem. Otherwise you just nail mating channels on door and jamb, following installation directions. Note that there are male and female channel strips, and that the male strip mounts on the door at the top and along the lock side, but it goes on the jamb along the hinge side of the door.

J-strips will require removal of the door from its hinges. Then grooves are routed or cut into the edge of the door to accept one half of the interlocking strips. The other half of the strip mounts on the jamb.

To weatherstrip the bottom of a door, you have two basic choices: sweeps and threshold strips.

Door sweeps are easy to install and they'll do a good job if you keep them properly adjusted. Usually they consist of a rubber flap mounted in an aluminum channel. You screw the channel to the inside of the door so that the flap butts up snugly against the threshold when the door is closed. The aluminum mount usually has slots for the mounting screws so you can adjust the height of the stripping to compensate for wear.

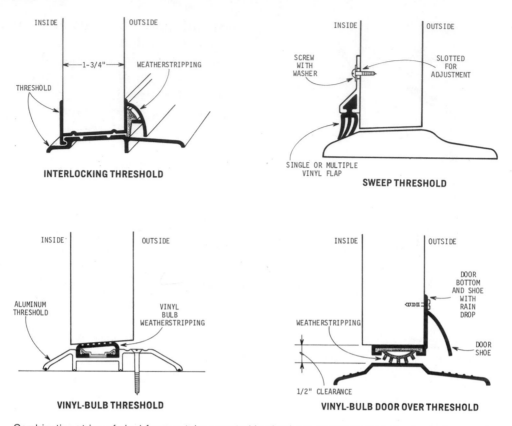

Combination strips of vinyl foam or tube mounted in aluminum cost more than simple vinyl-covered foam stripping, but they are better on warped doors, and they stay in place better. Foam strips often go up with thumb tacks that tend to pull out. Combination strips are secured with nails. To attach them, close the door, press the vinyl edge against the door—nailing through the aluminum backing strip. Door bottoms are tricky stripping jobs because of the friction of door against threshold. The interlocking threshold type provides a tight, long-lasting seal, but it's hard to install. Door sweeps are cheaper and easier to install, but require adjustment as they wear. A vinyl bulb threshold is good where the old threshold is worn out; bulbs are replaceable if they wear. A vinyl bulb can also go on a door bottom, but you may need a new threshold of aluminum.

Sometimes sweeps will cause problems if your home is carpeted; the rubber flap hangs up on the carpet. There is an "automatic" sweep with a simple camlike device that lifts the sweep when the door is open to solve the carpet hangup problem.

Threshold-type weatherstripping comes in a variety of types. One features a rubber seal in an aluminum channel mount that fits the bottom of the door. To install this type you'll have to saw a certain amount off the bottom of the door to provide clearance for the seal. This type of door seal works well if you have a flat wooden threshold that hasn't been worn so much that a good seal is impossible. The rubber seal can wear, but it is replaceable.

Another type of door bottom weatherstripping puts the rubber seal in the threshold, which is made of aluminum. You might use this one if you have no threshold at present, or if your current threshold is so worn it needs replacement. Most often this type of stripping is used in new construction, especially for basement and garage doors where the aluminum threshold is fastened directly to the concrete floor.

Interlocking threshold seals are the hardest to install, but they provide a good weatherseal. The main problem in installation is getting the perfect alignment required for proper meshing of the aluminum door and threshold parts. This type requires its own aluminum threshold, so unless yours is worn or nonexistent you'd be better off with a simple sweep in most cases.

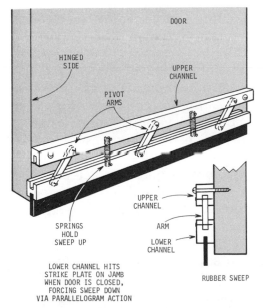

This automatic door sweep is spring loaded to rise above carpet level whenever you open the door. Closing the door forces the end of the sweep against a strike plate on the jamb, which forces the sweep downward, producing a tight seal between the threshold and the sweep.

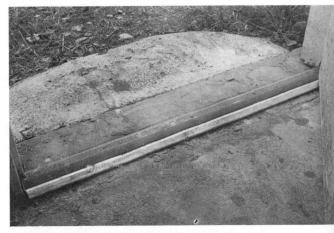

A vinyl bulb threshold is common on basement and garage doors where it is fastened directly to the concrete floor. The bulb is replaceable.

WINDOWS AND DOORS

WINDOWS

Windows lose a lot of heat. Even if yours are well weatherstripped, even if they're made of insulating glass (twin panes), even if they're fitted with storm windows, you can expect them to lose 6 to 10 times as much heat as an insulated wall of the same area. And if they're made with just a single layer of glass they may lose 12 to 20 times as much as an equal area of insulated wall. Obviously, windows play a very important part in the energy efficiency of your home. Improve the heat-retaining qualities of your windows, and you've done a lot to cut your heating costs. And of course you'll cut your cooling costs and improve summer comfort at the same time.

Six basic factors largely determine the energy efficiency of your windows: size, construction, location, number of glass layers, exterior shading, interior improvements. (For detailed instructions on computing heat loss through various windows and other parts of your home, see the explanation beginning on page 53.)

Let's look at energy efficiency factors one at a time and see how we can manipulate them for optimum energy savings.

Window Size. Obviously, the larger a window, the more heat it will lose. A window 24 square feet in area will lose twice as much heat as one just 12 square feet in area. As a general rule, then, keeping windows small will help keep heating and cooling bills small. If you like the feeling of space and light you get from large windows, all is not lost. As we'll see later, you can get away with large windows in certain locations.

Window Construction. The material used for making the window's frame or sash has an effect on the window's insulating quality. Wood is the best material, or wood covered with vinyl. Metal conducts heat much more readily than wood, so you can figure an aluminum sash will lose more heat than a wooden one. Tests show that metal sash construction loses five to 10 percent more heat than wood construction. So when you have a choice, lean toward wood. If you like the low maintenance of aluminum, consider vinyl-clad or aluminum-clad wood — also virtually maintenance free. A further advantage of wood over aluminum is the fact that wood windows are much less likely to sweat in winter than aluminum windows are. New designs in aluminum windows using plastic thermal breaks are solving these problems. The Architectural Aluminum Manufacturers Association claims these designs perform as well as wood. But test data are scarce.

Location. It's a happy fact that windows that face the sun act as solar collectors. As a general rule, a single-glazed window facing south will score even during a sunny day. That

Aluminum combination storm windows are low-cost and easy to install. This unit goes up with a few screws and some caulk. Such combination storm windows are often available for around $15.

is, it will gain roughly as much heat during the hours of sunshine as it will lose all day and all night. If the window is double glazed it will probably gain twice as much heat as it loses. Windows facing south southeast or south southwest will do just about as well as those facing directly south. Southeast or southwest windows will fare about 30 percent worse. And you can forget about collecting solar energy in useful amounts from north, east, and west windows.

Of course, not every winter day is sunny. If you live in an area where more than half your winter days are cloudy, even double-glazed south windows will wind up as net losers of energy. But if you do want large windows, put them on your south, southeast or southwest walls.

Number of Glass Layers. Another way to improve the efficiency of a window is to double or triple glaze. The insulating value of a window is almost exactly proportional to the number of glass layers in that window. Twin-glass windows lose half as much heat as those with just a single layer of glass. So do windows with storm sash. Add a third layer and the window will lose about a third as much heat as a single-glazed window.

Surprisingly, a sheet of clear plastic is as good an insulator as a sheet of glass. So a temporary storm window made with plastic film will do the same job as a glass storm window. Thus, if you're primarily interested in saving money, plastic film is your an-swer. Your next best choice is do-it-yourself glass storm windows. These cost more than plastic film but less than commercial storm windows. The most expensive option is to replace a single-glazed window with a new double-glazed unit. To determine the relative costs and savings of each of these possibilities, see the heat-loss calculations near the end of this chapter.

Exterior Shading. This can improve a window's energy efficiency two ways. Proper shading can keep the sun from hitting a window in the summer, thus cutting heat gain and keeping your home cool. This same shading device can allow the sun to hit the window in the winter, thus allowing the window to provide you with some free solar heat. The shading device can take the form of deciduous trees, or awnings, or a roof overhang. The leaves of the deciduous tree provide the summer shading and then drop off, allowing the winter sun to penetrate. Awnings and overhangs do their jobs only if sun angles are properly utilized. Since a summer sun is high in the sky, and a winter sun quite low, an awning or overhang of the right size will be able to block summer sunlight, yet won't get in the way of winter sunlight. You'll find more on this in Chapter 7.

Interior Improvements. A variety of improvements you can make inside your home will also upgrade the energy efficiency of your windows. Shades,

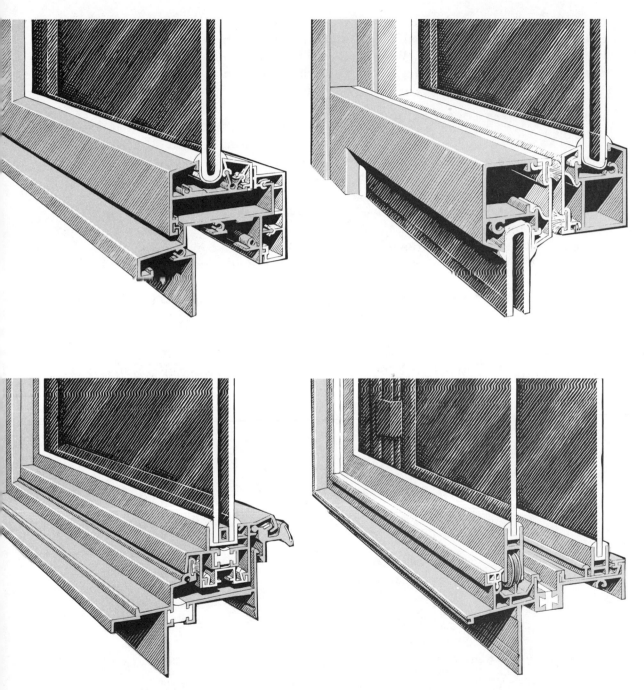

High-quality, double-glazed aluminum windows feature plastic thermal breaks or plastic cladding to cut down on heat flow and condensation. Shown here are Alcoa casement and double-hung windows (top row) and Reynolds and Capitol sliding windows (bottom row). *Drawings by Ray Pioch.*

draperies, insulating shutters, and heat-absorbing surfaces can all be put to work to cut heating costs. We'll go over these in detail later on.

HOW TO CUT HEAT LOSS THROUGH WINDOWS

A lot of modern homes have big picture windows. When they were built, there was a lot of wishful thinking about how much solar heat these windows could trap. Solar windows, they were called. And solar windows they can be if they're located not much more than 45 degrees east or west of south, and if they aren't blocked by trees or other obstructions during the

daylight hours. But since they lose almost as much heat when the sun is down (or clouded over) as they gain when the sun is out, they rarely do more than break even.

A number of simple measures can help them do better than break even, however. Maybe you already have the means to help these windows save energy. A set of lined draperies can cut heat losses by about 30 percent—if you remember to close them during the hours of darkness. Folding foam screens can do even better. Those shown here can cut heat losses by 75 percent whenever they're closed. And they needn't look ugly or utilitarian if you use your imagination

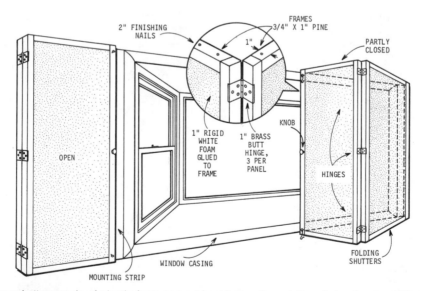

Window shutters made of plastic foam can cut heat losses through the window by about 75 percent whenever they're closed. Glue each foam panel into a pine frame. The frame serves only to hold the screws for the hinges and the other hardware, which would pull out of the foam. Size the shutter to overlap sides of the window trim by about three inches and mount each shutter on a mounting strip cut to the same thickness as the window casing. This will allow the shutter to fit snugly against window casing whenever the shutter is closed.

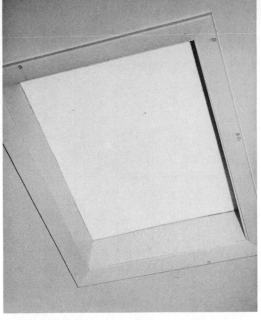

A piece of plastic foam cut to fit snugly inside a skylight shaft is a cheap means of reducing heat loss at night. Just slip the foam into the opening; it's lightweight enough to stay in place without support.

in decorating them. Cover each of the panels with wallpaper, maps, or inexpensive prints. Use decorative brass hardware.

Each panel is made of one-inch plastic foam surrounded by a frame of pine. The only purpose of the frame is to provide secure anchorage for the hinges. Make each panel about a foot wide. This size is easy to handle, and it stores neatly against the wall when the screen is open during the day. Alter the width of the panels, and the number of panels you use in order to cover the window opening and sash with about three inches of overlap at the ends.

The screens hinge-mount to mounting strips, fastened with Molly bolts to the wall. These strips should be the same thickness as the window casing and about 1¾-inches wide. If your window has a sill, size the screen to rest on the sill when closed. Glue a strip of felt to the bottom of the screen to protect the sill against possible scratching or marring.

These screens, fitted to a south-facing window, can turn it into a decent solar heater. Example: A four-by-eight-foot window can collect about 44,000 Btu's on a sunny day. If you close the screen at night you can expect to retain almost 32,000 of those Btu's. This works out to about 13 cents a day (assuming you live at approximately 40 degrees latitude, that the outside temperature is 20 degrees F, and that you heat with fuel oil at 40 cents a gallon.)

The same type of foam board material used for shutters can also be used to insulate skylights. Just cut the foam to the size of the skylight opening and press it into place. The foam is so light it will stay in position as long as it fits the skylight opening snugly. This simple arrangement will cut nighttime heat losses about 75 percent. Of course the foam blocks light, so you won't want to leave it in place during the day, but there are a couple of ways to cut heat losses through skylights during the day, without losing their ability to provide light.

One trick is to cut a sheet of .080-inch clear acrylic (about $1.20 a square foot) to fit into the skylight opening. If your skylight has snap-in-snap-out screens you can just remove the screen, place the sheet of plastic on top of the screen, then snap the screen back in place. It will hold the plastic in position, and the plastic will cut heat losses 30 to 50 percent. If your skylight has no screen, just nail a simple frame of quarter round molding around the inside of the sky-light opening. Then you can slip the sheet of plastic up through the opening and rest it on top of the quarter round frame. You could use glass for this job, but plastic is safer, and it's a slightly better insulator.

I've also used clear acrylic plastic to double glaze the window in my back door. Just cut the sheet two inches larger than the area of the window opening, then screw it directly to the inside face of the door. I use lightweight .080-inch plastic, but if you use heavier stock (¼-inch will do) you can burglar-proof your door at the same time. Acrylic this thick is just about unbreakable and it will prevent an intruder from knocking out a pane of glass, reaching in and unlocking the door.

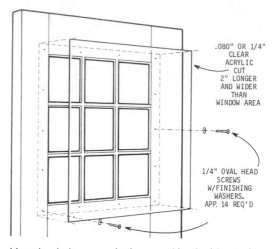

Many back doors can be improved by double glazing the window area as shown here. Use .080-inch acrylic for economy, or .25-inch acrylic to burglar proof the door. The thicker grade is virtually unbreakable. If well anchored, it prevents a burglar from knocking through the opening to reach the door's lockset. Put the plastic on the inside of the door.

Conventional Storm Windows. If you want to install storm windows, you have two choices. You can go with simple single-pane storms that cover the entire window area with a single piece of glass. They usually hang in place, can't be opened, have no screens. If you have airconditioning or for some other reason never want to open your windows, you can leave these in place year-round, saving on both heating and cooling costs. Otherwise you put them up in the fall, and take them down in the spring, replacing them with screens. This can be a bit troublesome, at least on second- or third-floor windows. But this type of storm window is cheap—usually about 10 bucks for the average window—and if you can leave it in place all year, it's a good choice.

Combination Storm Windows. The type that open and close and come with screens—can cost anywhere from $15 to $30 for the average window, depending on overall quality of construction and materials. Though they cost more than the simple single-pane type, they're more versatile, and a lot easier to live with. And they don't require you to invest in separate screens. So overall, they're a better bargain than they seem at first. On the other hand, they're not necessary if you always keep your windows closed. Single pane storms will do the job for less money. My advice: Use the combination type if you open your windows in the summer. Use the single-pane type if you don't. Whichever type you choose, mounting is pretty easy, as long as the windows fit right. Usually it's just a matter of applying some caulk and driving a few screws.

Both combination and single-pane storm windows you buy commercially are going to be framed in aluminum. Wood-framed storm windows are slightly better insulators. Without too much trouble you can make your own wood-framed storm windows, and save money two ways: They'll cost less to make, and they'll insulate better once they're made. If you do make your own, though, you ought to stick to simple storms. Combination units with storms and screens are too much bother to make yourself.

MAKING YOUR OWN STORMS

If you do go to the bother of making your own storm windows, you might consider double glazing them. They won't be much harder to make with double glazing than they are with single, and your only extra expense will be for the second sheet of glass. Double-glazed storms don't make much sense in warm climates, but they'll pay off in chilly northern climates.

Here's how to make a double glazed window. (If you decide to single glaze, the same techniques are used; simply use a single sheet of glass instead of the two-sheet sandwich described here.)

Start with the frame. Use clear, dry pine to avoid subsequent problems with warping and twisting. Use half-

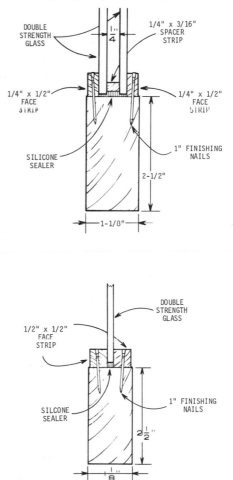

You can make your own storm sash—either double or single-glazed—using clear pine and double-strength window glass. Use hook-type hardware to hang the windows in place. Take them down for summer and replace them with screens, or, if you aircondition, leave them up all year.

lap joints at the corners for strength, sizing the frames to fit inside the exterior trim of the window you're weatherproofing. Use a waterproof glue (epoxy or a resorcinol) and one-inch brads to secure the joints. Glue and nail on one of the face stops.

Glass for the window should be sized 1/4-inch shorter and narrower than the inside measurements of the frame. This provides 1/8-inch edge clearance all around to allow for thermal expansion of the glass. Carefully clean the inside surfaces of the two panes of glass. Then, using silicone sealer and spacer-strip separators, cement the two sheets together. Set the strip in about 1/16-inch from the edges of the glass, and fill the resulting channel with silicone to provide a good seal. Now let the silicone and the frame set for a day.

When silicone and glue have cured, place the glass in the frame and position it with the proper 1/8-inch clearance all around. Work silicone into this gap with a narrow putty knife, and nail on the other face strip. Prime with an exterior house-paint primer, apply finish coat, screw on the hanging hardware, and install the window. If you don't plan to remove the window, you can fasten it in place with finishing nails. A bead of caulk laid around the mating surfaces of window trim and storm window will provide 100-percent sealing against drafts if you elect to nail the window up. If you use hanging hardware you can seal with caulking cord; this stuff will peel away when you remove the storm window in the spring. Replace it with fresh cord every fall.

BEADWALL

The most innovative, energy-efficient form of window I've ever seen is called Beadwall. It was invented by Steve Harrison. Plans and licenses for its construction are being marketed through Zomeworks. Essentially, Beadwall is a double-glazed window equipped with a device that can blow Styrofoam beads into the space between the two panes of glass whenever extra insulation is desired. This same device can also suck the beads

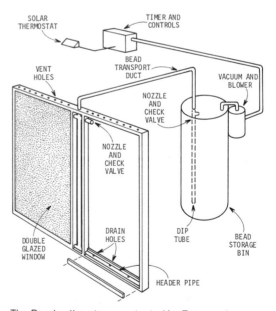

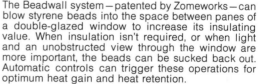

The Beadwall system—patented by Zomeworks—can blow styrene beads into the space between panes of a double-glazed window to increase its insulating value. When insulation isn't required, or when light and an unobstructed view through the window are more important, the beads can be sucked back out. Automatic controls can trigger these operations for optimum heat gain and heat retention.

back out of the window whenever transmission of light is more important than insulation. A window of this type, with a 3½-inch space between the panes might have an R rating of around 11, far higher than any other type of window, and as good as the walls in many homes.

The possibilities of Beadwall are almost unlimited. You could make the entire south wall of a home with Beadwall. Whenever you want to capture solar heat, or let in light, the beads are sucked out of the window. During cloudy periods, and during the night, you blow the window full of beads and get great insulation. Skyscrapers could be made of the stuff, and so could greenhouses. (Zomeworks sells plans for greenhouses.)

Unfortunately, there are drawbacks. The equipment for blowing and storing the beads can get expensive. Tempered glass is required to withstand the pressures generated when the beads are blown in and out. You can't cut tempered glass so you have to order it in one of its stock sizes. Storage for the beads in 55 gallon drums can get bulky. Commerical development of a complete Beadwall system could bring the price down significantly, but right now, you'll probably have to spend at least a couple hundred bucks for even a small do-it-your installation. Fortunately, large windows cost much less per foot than small ones, and Beadwall makes most sense when used big. If you're interested in Beadwall, write Zomeworks. In addition to their plans they also sell blower-and-drum units for storage and blowing of the beads. The address: Box 712, Albuquerque, NM 87103.

REDUCING HEAT GAIN THROUGH WINDOWS

Windows not only lose a lot of heat to the outdoors during cold weather, they can also admit so much heat in the form of solar radiation during the warm season that they seriously overheat your home. The problem gets so bad in the south that many homes are equipped with windows of reflective, mirrorlike glass. These still admit light, but they also reflect a great deal of light. And as a result, they prevent significant amounts of heat from entering the house.

Short of installing special glass in your windows, there are a few things you can do to cut down on heat gain. The best route is to cut off the sunlight before it gets into the house. Curtains, shades, and blinds are better than nothing, especially if they are light in color. They can reflect quite a bit of light back out the window before it can be converted to long-wave infrared radiation. (This happens when the sun's short-wave IR strikes a dark surface.) Long-wave IR can't pass through glass, so once it forms inside your house, it's trapped. Blinds, draperies, and curtains also protect such things as furniture and carpets from the bleaching effects of direct sunlight. Still, you're better off

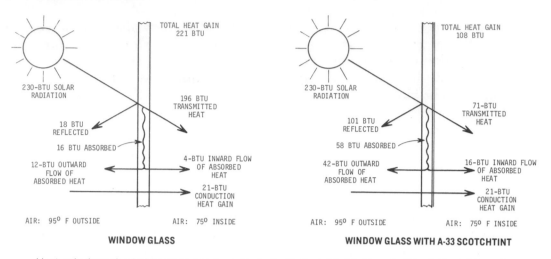

Heat gain through window glass can be cut in half with Scotchtint light control film in the A-33 grade. With A-18 grade, total transmission is cut to only 70 Btu per hour per square foot of window area.

intercepting the sun outside your home. (For a review of awnings and roof overhangs designed to admit winter sun and block summer sun, see the heading "Exterior Shading," earlier in this chapter.) Another approach to sun control is aluminized polyester film. You can apply this material to the inside of your windows and cut solar heat gain by about half.

This film is made by 3M under the name Scotchtint. With the Scotchtint in place, you'll find your windows are mirrorlike from the outside, making it difficult to see in. But you can see out with no problems. This privacy factor can be a boon to those thousands of homes with big picture windows that open up right onto the street.

Scotchtint comes in two grades. A-33, the stuff that cuts transmission in half, is sold by Sears. The other grade, A-18, cuts heat gain by about 70 percent, but it's available only through franchised installers. Interestingly, Scotchtint not only keeps outside heat from getting in, it also traps heat already inside your home. As a result, you can leave the stuff on your windows year-round. Your windows will lose most of their ability to collect heat from the sun, but they'll make up for this by losing less heat both day and night.

Applying the film is pretty simple. First of all you have to make sure the window is clean. Dust and dirt particles trapped between film and window are a problem, so clean with ammonia (clear, not sudsy) and dry with a squeegee. Don't use towels. Cut a piece of the film and its protective backing a bit oversize for the window you want to cover. Wet the window down with water, and do the same for the film. Place the film in position on the window and smooth out the wrinkles. Remove the protective backing sheet, wet the film again and squeegee it smooth. Using a razor blade and a straightedge, trim the film, leaving a margin of about 1/16-inch all around. Rewet the film, squeegee again, and finally mop up the excess water with a soft cloth.

DOORS

If your windows are your home's biggest heat wasters, your doors probably come in a close second. The average solid wood door has an R-value of around 2. Compare that to the R-value of a typical insulated frame wall (about 14) and you can see that your door loses about seven times as much heat per square foot as your insulated walls. And these figures don't take into account any air infiltration around the door. If you add in the infiltration factors, your doors will fare even worse. Of course weatherstripping can cut down on the infiltration problem, but if you really want to lower heat losses through your entryways, you'll have to modify the doors themselves.

Storms or Beefed-up Insulation? Storm doors are one solution. A wood storm door with a glass area of about 50 percent will raise your R-value to about 3.7. Not spectacular, but an improvement. Another good idea is to insulate your door. You can do the job fairly

simply with one-inch foam plastic. The simplest route is to cover just the center part of your door, keeping the insulation inside the doorknob and other hardware, and stopping it short of the edge of the door. This way the additional thickness of the insulation won't interfere with hardware and the operation of the door.

You can cement the foam to the door, then cover it with hardboard for better durability and appearance.

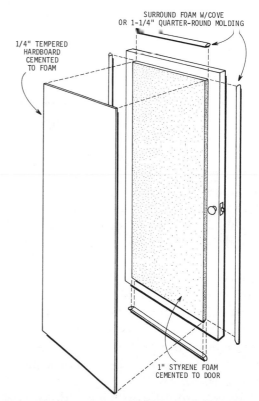

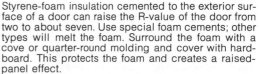

Styrene-foam insulation cemented to the exterior surface of a door can raise the R-value of the door from two to about seven. Use special foam cements; other types will melt the foam. Surround the foam with a cove or quarter-round molding and cover with hardboard. This protects the foam and creates a raised-panel effect.

Run molding around the edges of the insulation and you'll have an attractive raised panel effect. A single sheet of foam in the one-inch thickness will raise the R-value of your door up to around seven or so. For best effectiveness, put the foam on the outside of the door. Placing it inside may violate fire codes in some localities, unless the foam is covered with a layer of fireguard gypsum board (not a particularly durable material for a door covering).

Foam-core Steel Doors   If your door is old and in bad shape you might think about replacing it with one of the new foam-core steel doors. Early foam-steel doors were not much better energy savers than wood doors because the steel skin was one piece. As a result the metal skin conducted heat out around the foam insulation. New designs have solved that problem by interrupting the steel skin around the edges of the door, forming a thermal break. In addition, steel doors can utilize magnetic weatherstripping of the type used on modern refrigerators and freezers. A flexible magnet inside a plastic sleeve of weatherstripping pulls the weatherstripping tight against the door whenever it is closed to provide an excellent seal against air infiltration.

If you decide on a new door you have two ways to go: You can simply buy a new door and hang it in your old jamb, or you can go for a prehung door, complete with its own frame, weatherstripping, hinges and hard-

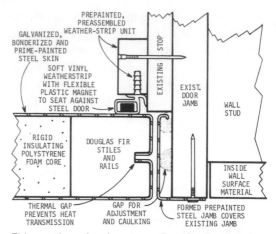

This section of a foam-core Ever/Strait steel door shows the thermal break in steel skin and installation details. This is a door specifically designed for easy installation in remodeling situations. It's slightly undersize to fit into the old opening. Note the freezer-type magnetic weatherstripping.

ware. If you only want to change the door without any kind of an overall facelift, the simple door is the easy way to go. But if you are interested in remodeling your entrance as well as raising its R-value, the prehung door is the better way to go. To install it, remove your old door along with its casing and hinges. The new prehung door then fits inside the old opening (special remodeling doors are sized a bit smaller than your original door to make this fit possible) and gets nailed in place, caulked, and weatherstripped.

AIRLOCK ENTRYWAYS

Insulating your door, adding a storm door, or installing an energy-efficient door all help cut losses while the door is closed, but none of these ideas helps at all whenever you open the door to enter or leave your home. Every time the door swings open, in comes a blast of cold air. The solution here is an airlock—a sort of enclosed porch with its own door. This arrangement lets you keep your heat, and also provides a welcome bonus: It's a handy, out-of-the-weather spot to remove boots and snow-covered coats before you step inside.

If you already have a front stoop, you can build an airlock right on top of it without much trouble. Just nail a sill of 2x4s to the stoop with masonry nails, frame the walls with more 2x4s, sheath with plywood and cover with siding to match your home. You can use a simple prehung storm door for the airlock door. (See Chapter 7 for more details.)

If your home has a canopy over the door, or a wide overhanging soffit, just build the airlock so that the canopy or soffit forms the roof. Otherwise, frame the canopy roof with 2x4s, sheath with plywood, and shingle to match your main roof. When it's finished, your airlock will not only cut down on drafts into your home, but it will also reduce heat losses through the door, serving as a third storm door.

HOW TO CALCULATE HEAT LOSS
(through windows and other parts of your home)

To find yearly heat loss through a window, multiply figures in the table below by the area of the window in square feet. Multiply the result by 24 (number of hours in a day), and multi-

ply that result by the annual degree day figure for the city closest to your home (shown in the table on the next two pages). Your final result will be Btu's lost through that window per year.

Window type	COEFFICIENT OF TRANSMISSION (U) Btu per hour/square foot/degree F
Single glazing	1.13
Double glazing	.58
Triple glazing	.36

COST OF FUEL IN DOLLARS PER THOUSAND BTU

How much does it cost you per year to replace the heat that goes out through the window? To find out, simply multiply the yearly Btu-loss figure you computed above by the price you pay per Btu of heat. The table below shows costs of basic fuels by quantity and by costs per Btu:

Electricity	$ per kwh	$ per thousand Btu
	.02	.0059
	.025	.0074
	.03	.0088
	.035	.0103
	.04	.0118
	.045	.0132
	.05	.0147
Fuel oil	$ per gallon	$ per thousand Btu
	.25	.0025
	.30	.0030
	.35	.0035
	.40	.0040
	.45	.0045
	.50	.0050
	.55	.0055
Natural gas	$ per 1000 cu. ft.	$ per thousand Btu
	1.50	.0022
	1.75	.0025
	2.00	.0029
	2.25	.0032
	2.50	.0036
	2.75	.0039
	3.00	.0043
	3.25	.0046

HEATING DEGREE DAYS

Heating degree days indicate the number of degrees the daily average temperatures are below 65° F. The figure 65° is used because it is the average outdoor temperature below which most people feel the need for indoor heat. Degree days are computed like this: A day with an average temperature of 50° has 15 heating degree days (65 - 50 = 15). Days that average 65° or over have no heating degree days.

This table shows the monthly and annual totals by city. Note: There is a direct relation between heating degree days and fuel consumption that allows fairly accurate comparisons of heating energy needs. For example, it would take roughly seven times as much energy to heat a house in Duluth as it would for the same house in New Orleans (10,000 ÷ 1385 = 7). And the relation of heating degree days to energy needs is fairly constant, whether any 100 degree days occur over many or only a few calendar days. The fuel industry uses this system on a month-to-month basis to determine fuel rates and peak demands. Table source: *Climatic Atlas of the United States*

NORMAL TOTAL HEATING DEGREE DAYS (Base 65°)

STATE AND STATION	JUL	AUG	SEP	OCT	NOV	DEC	JAN	FEB	MAR	APR	MAY	JUN	ANNUAL
ALA. Birmingham	0	0	6	93	363	555	592	462	363	108	9	0	2551
Huntsville	0	0	12	127	426	663	694	557	434	138	19	0	3070
Mobile	0	0	0	22	213	357	415	300	211	42	0	0	1560
ALASKA Anchorage	245	291	516	930	1284	1572	1631	1316	1293	879	592	315	10864
Barrow	803	840	1035	1500	1971	2362	2517	2332	2468	1944	1445	957	20174
Fairbanks	171	332	642	1203	1833	2254	2359	1901	1739	1068	555	222	14279
Juneau	301	338	483	725	921	1135	1237	1070	1073	810	601	381	9075
Nome	481	496	693	1094	1455	1820	1879	1666	1770	1314	930	573	14171
ARIZ. Flagstaff	46	68	201	558	867	1073	1169	991	911	651	437	180	7152
Prescott	0	0	27	245	579	797	865	711	605	360	158	15	4362
Tucson	0	0	0	25	231	406	471	344	242	75	6	0	1800
ARK. Fort Smith	0	0	12	127	450	704	781	596	456	144	22	0	3292
Little Rock	0	0	9	127	465	716	756	577	434	126	9	0	3219
Texarkana	0	0	0	78	345	561	626	468	350	105	0	0	2533
CALIF. Eureka	270	257	258	329	414	499	546	470	505	438	372	285	4643
Fresno	0	0	0	78	339	558	586	406	319	150	56	0	2492
Mt. Shasta	25	34	123	406	696	902	983	784	738	525	347	159	5722
San Diego	6	0	15	37	123	251	313	249	202	123	84	36	1439
San Francisco	81	78	60	143	306	462	508	395	363	279	214	126	3015
COLO. Alamosa	65	99	279	639	1065	1420	1476	1162	1020	696	440	168	8529
Denver	6	9	117	428	819	1035	1132	938	887	558	288	66	6283
Pueblo	0	0	54	326	750	986	1085	871	772	429	174	15	5462
CONN. Bridgeport	0	0	66	307	615	986	1079	966	853	510	208	27	5617
Hartford	0	6	99	372	711	1119	1209	1061	899	495	177	24	6172
New Haven	0	12	87	347	648	1011	1097	991	871	543	245	45	5897
DEL. Wilmington	0	0	51	270	588	927	980	874	735	387	112	6	4930
FLA. Jacksonville	0	0	0	12	144	310	332	246	174	21	0	0	1239
Miami Beach	0	0	0	0	0	40	56	36	9	0	0	0	141
Tallahassee	0	0	0	28	198	360	375	286	202	36	0	0	1485
Tampa	0	0	0	0	60	171	202	148	102	0	0	0	683
GA Atlanta	0	0	18	127	414	626	639	529	437	168	25	0	2983
Columbus	0	0	0	87	333	543	552	434	338	96	0	0	2383
Savannah	0	0	0	47	246	437	437	353	254	45	0	0	1819
IDAHO Boise	0	0	132	415	792	1017	1113	854	722	438	245	81	5809
Idaho Falls 46W	16	34	270	623	1056	1370	1538	1249	1085	651	391	192	8475
ILL. Cairo	0	0	36	164	513	791	856	680	539	195	47	0	3821
Chicago	0	0	81	326	753	1113	1209	1044	890	480	211	48	6155
Springfield	0	0	72	291	696	1023	1135	935	769	354	136	18	5429
IND. Evansville	0	0	66	220	606	896	955	767	620	237	68	0	4435
Indianapolis	0	0	90	316	723	1051	1113	949	809	432	177	39	5699
South Bend	0	6	111	372	777	1125	1221	1070	933	525	239	60	6439
IOWA Des Moines	0	9	99	363	837	1231	1398	1165	967	489	211	39	6808
Dubuque	12	31	156	450	906	1287	1420	1204	1026	546	260	78	7376
KANS. Dodge City	0	0	33	251	666	939	1051	840	719	354	124	9	4986
Goodland	0	6	81	381	810	1073	1166	955	884	507	236	42	6141
KY. Covington	0	0	75	291	669	983	1035	893	756	390	149	24	5265
Louisville	0	0	54	248	609	890	930	818	682	315	105	9	4660
LA. Baton Rouge	0	0	0	31	216	369	409	294	208	33	0	0	1560
New Orleans	0	0	0	19	192	322	363	258	192	39	0	0	1385
Shreveport	0	0	0	47	297	477	552	426	304	81	0	0	2184
MAINE Caribou	78	115	336	682	1044	1535	1690	1470	1308	858	468	183	9767
Portland	12	53	195	508	807	1215	1339	1182	1042	675	372	111	7511
MD. Baltimore	0	0	48	264	585	905	936	820	679	327	90	0	4654
Frederick	0	0	66	307	624	955	995	876	741	384	127	12	5087
MASS. Blue Hill Obsy	0	22	108	381	690	1085	1178	1053	936	579	267	69	6368
Boston	0	9	60	316	603	893	1088	972	846	513	208	36	5634
Nantucket	12	22	93	332	573	896	992	941	896	621	384	129	5891
MICH. Detroit (City)	0	0	87	360	738	1088	1181	1058	936	522	220	42	6232

NORMAL TOTAL HEATING DEGREE DAYS (Base 65°)

STATE AND STATION	JUL	AUG	SEP	OCT	NOV	DEC	JAN	FEB	MAR	APR	MAY	JUN	ANNUAL
Grand Rapids	9	28	135	434	804	1147	1259	1134	1011	579	279	75	6894
Marquette	59	81	240	527	936	1268	1411	1268	1187	771	468	177	8393
Sault Ste. Marie	96	105	279	580	951	1367	1525	1380	1277	810	477	201	9048
MINN. Duluth	71	109	330	632	1131	1581	1745	1518	1355	840	490	198	10000
International Falls	71	112	363	701	1236	1724	1919	1621	1414	828	443	174	10606
Minneapolis	22	31	189	505	1014	1454	1631	1380	1166	621	288	81	8382
Rochester	25	34	186	474	1005	1438	1593	1366	1150	630	301	93	8295
MISS. Jackson	0	0	0	65	315	502	546	414	310	87	0	0	2239
Vicksburg	0	0	0	53	279	462	512	384	282	69	0	0	2041
MO. Columbia	0	0	54	251	651	967	1076	874	716	324	121	12	5046
Kansas	0	0	39	220	612	905	1032	818	682	294	109	0	4711
St. Joseph	0	6	60	285	708	1039	1172	949	769	348	133	15	5484
Springfield	0	0	45	223	600	877	973	781	660	291	105	6	4561
MONT. Great Falls	28	53	258	543	921	1169	1349	1154	1063	642	384	186	7750
Havre	28	53	306	595	1065	1367	1584	1364	1181	657	338	162	8700
Missoula	34	74	303	651	1035	1287	1420	1120	970	621	391	219	8125
NEBR. Lincoln	0	6	75	301	726	1066	1237	1016	834	402	171	30	5864
Omaha	0	12	105	357	828	1175	1355	1126	939	465	208	42	6612
Valentine	9	12	165	493	942	1237	1395	1176	1045	579	288	84	7425
NEV. Ely	28	43	234	592	939	1184	1308	1075	977	672	456	225	7733
Las Vegas	0	0	0	78	387	617	688	487	335	111	6	0	2709
Reno	43	87	204	490	801	1026	1073	823	729	510	357	189	6332
N. H. Concord	6	50	177	505	822	1240	1358	1184	1032	636	298	75	7383
Mt. Wash. Obsy.	493	536	720	1057	1341	1742	1820	1663	1652	1260	930	603	13817
N. J. Atlantic City	0	0	39	251	549	880	936	848	741	420	133	15	4812
Trenton	0	0	57	264	576	924	989	885	753	399	121	12	4980
N. MEX. Albuquerque	0	0	12	229	642	868	930	703	595	288	81	0	4348
Silver City	0	0	6	183	525	729	791	605	518	261	87	0	3705
N. Y. Albany	0	19	138	440	777	1194	1311	1156	992	564	239	45	6875
Buffalo	19	37	141	440	777	1156	1256	1145	1039	645	329	78	7062
Central Park	0	0	30	233	540	902	986	885	760	408	118	9	4871
Syracuse	6	28	132	415	744	1153	1271	1140	1004	570	248	45	6756
N. C. Nasheville	0	0	48	245	555	775	784	683	592	273	87	0	4042
Cape Hatteras	0	0	0	78	273	521	580	518	440	177	25	0	2612
Raleigh	0	0	21	164	450	716	725	616	487	180	34	0	3393
Wilmington	0	0	0	74	291	521	546	462	357	96	0	0	2347
N. DAK. Bismarck	34	28	222	577	1083	1463	1708	1442	1203	645	329	117	8851
Devils Lake	40	53	273	642	1191	1634	1872	1579	1345	753	381	138	9901
Fargo	28	37	219	574	1107	1569	1789	1520	1260	690	332	99	9226
OHIO Akron	0	9	96	381	726	1070	1138	1016	871	489	202	30	6037
Cincinnati	0	0	54	248	612	921	970	837	701	336	118	9	4806
Cleveland	9	25	105	384	738	1088	1159	1047	918	552	260	66	6351
OKLA. Oklahoma City	0	0	15	164	498	766	868	664	527	189	34	0	3725
Tulsa	0	0	18	158	522	787	893	683	539	213	47	0	3860
OREG. Eugene	34	34	129	366	585	719	803	627	589	426	279	135	4726
Medford	0	0	78	372	678	871	918	697	642	432	242	78	5008
Portland	25	28	114	335	597	735	825	644	586	396	245	105	4635
PA. Erie	0	25	102	391	714	1063	1169	1081	973	585	288	60	6451
Philadelphia	0	0	60	291	621	964	1014	890	744	390	115	12	5101
Scranton	0	19	132	434	762	1104	1156	1028	893	498	195	33	6254
R. I. Block Is.	0	16	78	307	594	902	1020	955	877	612	344	99	5804
Providence	0	16	96	372	660	1023	1110	988	868	534	236	51	5954
S. C. Charleston	0	0	0	59	282	471	487	389	291	54	0	0	2033
Spartanburg	0	0	15	130	417	667	663	560	453	144	25	0	3074
S. DAK. Huron	9	12	165	508	1014	1432	1628	1355	1125	600	288	87	8223
Rapid City	22	12	165	481	897	1172	1333	1145	1051	615	326	126	7345
TENN. Bristol	0	0	51	236	573	828	828	700	598	261	68	0	4143
Chattanooga	0	0	18	143	468	698	722	577	453	150	25	0	3254
TEX. Amarillo	0	0	18	205	570	797	877	664	546	252	56	0	3985
Austin	0	0	0	31	225	388	468	325	223	51	0	0	1711
Corpus Christi	0	0	0	0	120	220	291	174	109	0	0	0	914
Fort Worth	0	0	0	65	324	536	614	448	319	99	0	0	2405
Houston	0	0	0	6	183	307	384	288	192	36	0	0	1396
UTAH Milford	0	0	99	443	867	1141	1252	988	822	519	279	87	6497
Wendover	0	0	48	372	822	1091	1178	902	729	408	177	51	5778
VT. Burlington	28	65	207	539	891	1349	1513	1333	1187	714	353	90	8269
VA. Cape Henry	0	0	0	112	360	645	694	633	536	246	53	0	3279
Lynchburg	0	0	51	223	540	822	849	731	605	267	78	0	4166
Norfolk	0	0	0	136	408	698	738	655	533	216	37	0	3421
WASH. Olympia	68	71	198	422	636	753	834	675	645	450	307	177	5236
Seattle	50	47	129	329	543	657	738	599	577	396	242	117	4424
Spokane	9	25	168	493	879	1082	1231	980	834	531	288	135	6655
Stampede Pass	273	291	393	701	1008	1178	1287	1075	1085	855	654	483	9283
W. VA. Charleston	0	0	63	254	591	865	880	770	648	300	96	9	4476
Elkins	9	25	135	400	729	992	1008	896	791	444	198	48	5675
Parkersburg	0	0	60	264	606	905	942	826	691	339	115	6	4754
WIS. Green Bay	28	50	174	484	924	1333	1494	1313	1141	654	335	99	8029
La Crosse	12	19	153	437	924	1339	1504	1277	1070	540	245	69	7589
Madison	25	40	174	474	930	1330	1473	1274	1113	618	310	102	7863
Milwaukee	43	47	174	471	876	1252	1376	1193	1054	642	372	135	7635
WYO. Cheyenne	19	31	210	543	924	1101	1228	1056	1011	672	381	102	7278
Lander	6	19	204	555	1020	1299	1417	1145	1017	654	381	153	7870
Sherian	25	31	219	539	948	1200	1355	1154	1054	642	366	150	7683

R AND U VALUES
FOR TYPICAL BUILDING MATERIALS

Wood bevel siding, ½"x8, lapped	R-0.81
Wood siding shingles,16", 7½" exposure	R-0.87
Asbestos-cement shingles	R-0.03
Stucco, per inch	R-0.20
Building paper	R-0.06
½" nail-base insul. board sheathing	R-1.14
½" insul. board sheathing, regular density	R-1.32
25/32" insul. board sheathing, regular density	R-2.04
¼" plywood	R-0.31
⅜" plywood	R-0.47
½" plywood	R-0.62
⅝" plywood	R-0.78
¼" hardboard	R-0.18
Softwood, per inch	R-1.25
Softwood board, ¾" thick	R-0.94
Concrete blocks, three oval cores	
Cinder aggregate, 4" thick	R-1.11
Cinder aggregate,12" thick	R-1.89
Cinder aggregate, 8" thick	R-1.72
Sand and gravel aggregate, 8" thick	R-1.11
Lightweight aggregate (expanded clay,	
shale, slag, pumice, etc.), 8" thick	R-2.00
Concrete blocks, two rectangular cores	
Sand and gravel aggregate, 8" thick	R-1.04
Lightweight aggregate, 8" thick	R-2.18
Common brick, per inch	R-0.20
Face brick, per inch	R-0.11
Sand-and-gravel concrete, per inch	R-0.08
Sand-and-gravel concrete, 8" thick	R-0.64
½" gypsumboard	R-0.45
⅝" gypsumboard	R-0.56
½" lightweight-aggregate gypsum plaster	R-0.32
25/32" hardwood finish flooring	R-0.68
Asphalt, linoleum, vinyl, or rubber floor tile	R-0.05
Carpet and fibrous pad	R-2.08
Carpet and foam rubber pad	R-1.23
Asphalt roof shingles	R-0.44
Wood roof shingles	R-0.94
⅜" built-up roof	R-0.33
Air Spaces (¾")	
Heat flow UP	
Non-reflective	R-0.87
Reflective, one surface	R-2.23
Heat flow DOWN	
Non-reflective	R-1.02
Reflective, one surface	R-3.55
Heat flow HORIZONTAL	
Non-reflective (also same for 4" thickness)	R-1.01
Reflective, one surface	R-3.48

Doors	
Solid wood 1" thick	U-.64 (.3 w/wood storm door)
Solid wood 1½" thick	U-.49 (.27 w/wood storm door)
Solid wood 2" thick	U-.43 (.24 w/wood storm door)
steel w/urethane core	U-.40
steel w/styrene core	U-.47

Is it worth upgrading the window? Going from single to double glazing will cut your window-heat-loss costs just about in half. Compare these sav-ings against the cost of your improvement and you have the answer.

Here's an example: Say you have a single-glazed window with an area of 12 square feet. You live in Bridgeport, Connecticut, and you heat with fuel oil at 40 cents a gallon. Check the first table in this section for the heat loss through single glazing. You find 1.13. Multiply this by the area of the window which is 12 square feet (1.13 × 12 = 13.56). Multiply the 13.56 by 24 (hours in a day) and you get 325.44. On the degree day table on the previous pages, you find that Bridgeport has an annual degree day rating of 5617. Multiply 325.44 × 5617 and you get 1,827,996 Btu's lost through that window per year. Multiply the Btu figure by the cost per Btu for fuel oil at 40 cents a gallon (.000004) and you get $7.31.

What kind of improvements would be worthwhile on this particular window? We know that going to double glazing will cut the heat loss—and the cost to replace the heat—by roughly half. You can prove this to yourself by going to the math above, starting with the heat loss figure for double glazing instead of for single. You'll get $3.75 as your cost per year to replace the heat passing through a double-glazed window. This means you'll save $3.56 a year by double glazing ($7.31 minus $3.75).

For about 50 cents you can add a plastic storm window made of polyethylene film with a lath frame. If you don't mind the appearance, you'll net a little over three bucks a year (the plastic needs annual replacement). A second choice would be to make, or have made, single-pane storm windows. One to cover our 12-foot window might cost $10—less if you make it yourself. This unit would pay for itself in a little more than two years. My local home center sells combination storm windows for $17. One of these would pay off in a bit less than five years.

That same home center also sells complete, vinyl-clad wood windows, weatherstripped, double-glazed and ready to install in place of the old single-glazed window in our example. Cost? Just about $100. Payoff time on that window would be 28 years, not including the cost of installation.

What about triple glazing? Usually, the only practical way to get triple glazing is to add some kind of storm window (plastic or glass) to a window that already has insulating glass. Running through the calculation on our hypothetical window, this time with triple glazing, we get a cost of $2.33 to replace heat lost through that window. This gives us a savings of $1.42 over our heating cost with double glazing. The costs for the storm window are the same as above. A plastic storm window would save a net of 92 cents a year. A single-pane storm would pay for itself in about seven years. A combination storm would pay off in 10½ years.

Note: The same formula used to check heat loss through windows can tell you the yearly heat loss through any other part of your home, too. All you need to know is the coefficient of transmission (U) for that part of your home, and its area in square feet. U values are noted at the bottom of the chart on the previous page. The formula then becomes:

$$U \times \text{area in square feet} \times 24 \times \text{annual degree days} = \text{yearly heat loss}$$

Where you are checking for heat loss through a single material, you can use U-values directly. If, however, you are calculating losses through more than one material (say a wall with siding, sheathing, insulation, and wallboard) add up all the R-values for those materials first. The reciprocal of that sum will be the U for the entire wall section. The table on the previous page gives the R-values you'll need to make your calculations.

Example: the U for a wall with lapped wood siding, ½-inch plywood sheathing, R-11 insulation, and ½-inch gypsum board would have an R-value of .81+.62+11+.45=12.88. The reciprocal or U would then be

$$\frac{1}{12.88} \text{ or } .076.$$

CONVENTIONAL HEATING SYSTEMS

Anyone with an ounce of sense has his car tuned now and then to keep it running at peak efficiency. The result is increased performance and more miles per gallon. In a short time the tuneup pays for itself in fuel savings.

Well, the same thing holds true for your home's heating system. A careful tuneup—much of which you can do yourself—will go a long way towards increasing efficiency. The steps you take depend on the type of heat you have. If you're stuck with electric heat, there's not much you can do in the way of a tuneup, though there are some energy-saving steps you can take. But if you heat with oil or gas, there's a lot you can do. And there's a lot of money to be saved along the way.

A list of things to be done—and a good order to follow when doing them—should include:

- Balancing the distribution system
- Cleaning, resetting, and possibly changing your thermostat.
- Cleaning and maintaining the furnace.
- Having the burner checked, adjusted, and possibly modified.
- Installing a humidifier.
- Insulating the air ducts.
- Providing combustion air.
- Converting to continuous-air circulation.

Let's go over these procedures in more detail.

BALANCING THE DISTRIBUTION SYSTEM

In a hot air system, heat is distributed throughout the house by means of air ducts. Hot water systems move heat via plumbing. Electric systems move it through wires. Your goal in balancing the system is to assure that just the right amount of heat—but no more than that—is delivered to every part of the house. In the hot air system you do this by means of dampers in the ducts. With hot water or hydronic systems you do it with valves in the plumbing lines. With electric heat you balance by means of the thermostats that control the resistance heating units.

Balancing is best achieved by trial and error. Heating experts do it with special instruments, but it's a hit and run affair—usually over in an hour. And instruments really have no sense of comfort. You are your own best judge as to what is the right temperature for a given part of the house.

For example, you may like a cool bedroom. The range and refrigerator in your kitchen may supply all the heat the kitchen needs. Your family room may require very little heat because your activities there keep you warm. Your living room may require extra heat to keep you warm while you sit reading, generating little body heat of your own. Upstairs rooms may need little heat since heat from the lower levels will rise there by convection. It's easy enough to see that different rooms require different temperatures. Balancing sees that they get it.

Electric baseboards often have their own thermostats, allowing for precise adjustment of the temperature in every room in the house.

Before you start balancing, check your registers and baseboard units to be sure they are not blocked by furniture, draperies, or the like. You want to allow free air movement, especially around baseboard units.

Electric heat is the easiest to balance. Usually each room will have its own thermostat. Sometimes, each heating unit will have its own 'stat. All you have to do is turn down the thermostats until you get the desired temperature level in each room. Don't overdo this by making big drops right at the start. Try a couple degrees at a time, and wait a day or two between drops. This allows the house to reach its new equilibrium, and it also eliminates the shock effect you'd get by dropping several degrees at once.

With forced hot air systems, adjustments are a bit trickier. If you're lucky, the ducts in your home will be fitted with dampers. Most hot air registers have dampers, built right in, but you'll get better results if you use dampers in the ducts. Go down to the basement and look at your ducting. Chances are there will be a few main trunk lines leading off from the furnace. Branching off these trunks will be smaller ducts serving individual rooms. Ideally, the control dampers should be in these branch ducts, right near the trunk. If there are no dampers you can install your own without too much trouble. Measure the diameter of the ducting and go to a heating supply outlet. Buy a damper of the correct size. It will look like a metal skillet the same diameter as the duct. To install it, slip the branch line loose from the trunk. Drill holes for the damper shaft opposite each other in the branch duct. Slip the damper in place, slide the damper handle on the shaft, and reconnect the duct. Finally run a strip of ducting tape around the joint, and the job is done. This shouldn't take more than a few minutes.

To adjust the dampers, simply close down on the ones that control the ducts to rooms that need the least heat. Most likely, the dampers will have little effect until they are nearly closed, so you can probably get the

Dampers in main heating ducts let you control how much heat goes to each room or zone in your home. This type of damper has a long handle aligned with the damper blade. When the handle points across the duct, the damper is shut.

job done fastest by overdamping at first, then slowly opening up if a day-or-two worth of experience indicates the room in question is too cold. Again, as for adjusting electric heat, allow adequate time to do the job right. It may take a week or two to arrive at the right setting for each room in your home. Once you find it, make a mark on the duct indicating the right position for the damper handle. This way you can reset the damper if it becomes moved accidentally, or if you also have central airconditioning and change your damper settings in the summer.

If you have no dampers in your ducts, and if you can't get at them to make your own installation, go to the dampers built into the registers in each room. Make your adjustments there. This isn't the ideal way to go, but you have no choice.

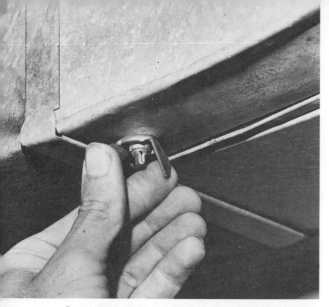

Some dampers have wingnut-type handles, and might be mistaken for hardware holding the duct together.

Hot water systems are adjusted just like hot air. Only difference is that you turn valves instead of dampers. Your home may not have valves in the hot water lines to your radiators. If not, make the adjustments at the valves right on the radiators.

Duct tape at all joints in a hot-air distribution system will eliminate air leaks and prevent heat losses. It pays to install duct insulation wherever ducts pass through unheated space.

THERMOSTAT KNOWHOW

After you've achieved good heating balance, the next step is to turn to your thermostat. First of all, is it in a good location? Ideally, a thermostat should be positioned where it won't be influenced by intermittent sources of heat or cold. Is yours located about four feet from the floor, and fairly near the center of the house? Is it out of direct sunlight? Is it located so it's not directly over a radiator or register? Away from an exterior wall or a large

Registers in many hot air systems have dampers built in. You should use these to balance your system only if there are no dampers built into the ducts. Duct dampers give better results.

window? Is it nowhere near a fireplace? If so, good.

If you've gone to the bother of balancing your heating as described above, it's also a good idea to keep the thermostat out of atypical rooms — those that you've throttled way down, for example — such as kitchens or bedrooms.

If the 'stat is poorly located, consider moving it to a better location. Once you're satisfied with location, consider the thermostat itself. Proper calibration and cleaning are important, but don't bother with this until you've decided whether you might want a new thermostat. Today you can buy a special clock thermostat that will automatically lower the heat at night, then raise it again before you get up the next morning. Over the years there has been a lot of argument about the value of nighttime setbacks, but a recent computer study has removed the doubt. Study figures show that savings up to 16 percent can result with 10-degree nighttime setbacks in mild climates (Los Angeles, Atlanta, and Dallas for example). Up north the percentage savings were not as high, but dollar savings were greater because the amount of fuel used — and as a result also saved — is greater in the north.

The thermostat is simple enough to install. It hooks up to the same two wires as your old one. No extra lines are required to run the clock. Price runs around $80 for a do-it-yourself installation kit for the Honeywell Chronotherm (Model T882A1120).

Dirty thermostats like this one can be cleaned with a soft brush. Heavy dust buildups can interfere with proper operation.

Note: Much of the effectiveness of nighttime thermostat setback will be lost if a rigid schedule is not followed. Once an automatic clock thermostat is set, it will perform all winter without attention. If you try to duplicate this performance by turning an ordinary thermostat up and down manually, you'll probably wind up saving yourself very little fuel in the long run unless you can be as faithful as a clock. Moral: If you decide to go with nighttime setback, get an automatic clock thermostat.

If you elect to stick with your old 'stat, be sure it's in good working order. Most modern units have a mercury switch, but some of the old contact-types are still in use. Remove the cover from your thermostat and you'll quickly see what kind you have. If it has contacts, run a strip of paper between the points to clean and burnish them. Dust off the works with a soft brush. If the thermostat is the mercury type, check to be sure it's level. A flat spot on the top or bottom of the 'stat is usually provided to check the level. If the thermostat is not level you can loosen its mounting screws and adjust it. Then retighten the screws.

While it's not particularly important for the thermometer on the thermostat to read out the actual temperature (as long as the thermostat shuts on and off consistently), you can adjust the thermometer to give accurate readings if you desire. Contact-type 'stats usually have a small allen screw for adjustments. With mercury-switch types you usually slide the mercury bulb up or down after loosening a tab. In either case, use a thermometer of known accuracy to calibrate the 'stat thermometer.

CLEANING AND
MAINTAINING THE FURANCE

Many fuel and furnace repair businesses offer service policies for furnaces. For a fixed amount of money per year (around $30 in my area) you get a yearly cleaning and tuneup, plus free service on anything that goes wrong with the furnace during the year. It's a good investment; just the cleanup and tuneup is worth that much. The free repairs — if required — are a bonus. It's an especially good investment if your furnace is getting on in years, and breakdowns are more common.

If you have a new furnace, and don't want to go with the service pol-

Here a limit control switches the fan on and off at preset temperatures. This one also has a switch to shut off the burner if the heat level gets too high. If you want to tune and set your own heating system, get a copy of your furnace manual. Service manuals are sold by Petroleum Marketing Education Foundation, Box 11187, Columbia, SC 11187; National Fuel Oil Institute, 60 E. 42 St., New York, NY 10017. Test instruments are available from Dwyer Instruments, Box 373, Michigan City, IN 46360.

icy, think twice. A yearly tuneup will save you anywhere from $25 to $65 every year in heating costs. So the service contract will probably pay for itself in fuel savings alone.

If you don't want a service policy or can't get one, you can still do most of the work yourself. This just requires the initiative to get involved with your heating unit's inner workings. To do things right you'll need a furnace manual to supply you with specifications. And you'll have to read and understand it. More about that later in this chapter.

Cleaning Comes First. Carbon buildups in combustion chambers and flue passages will cost you efficiency and money. The soot acts as an insulator, and a pretty efficient one at that. Figure on a 20-percent loss in heating efficiency with a deposit of soot around 1/4-inch thick. Getting rid of the soot isn't too difficult, just messy and unpleasant.

Start by shutting off all power to your furnace—either at the main switch, or at the fuse or circuit breaker that controls the furnace. The fuse or breaker will be at your main electrical service panel. Then remove the smoke pipe, and check the lining of the combustion chamber. It it's firebrick or solid steel, use a vacuum cleaner to clean it off. You'll get best results with no attachment on the end of the hose. Just scrape away with the end of the hose fitting.

If the lining of the combustion chamber is a fibrous clothlike material, leave it alone. It shouldn't need cleaning, and a heavy-handed scrubbing will only cause damage to the delicate material.

Flue passageways can be reached through the opening where they leave the furnace.

Now back to the furnace manual. If you have one, you can make a couple of adjustments, or at least check to see if the adjustments need to be made.

This applies to hot-air systems only. First thing to check is the temperature rise of the air going through the heat exchanger. Your furnace manual will tell you what the rise should be. To check it you simply measure the temperature of the air coming out of the furnace, and compare it to the temperature of the air coming in.

Drill a hole in each of the two chambers where ducts connect to the furnace. Stick a thermometer into one of these holes and take a reading after the furnace has been running for a few minutes. Write down your result and then check the temperature through the other hole. The difference between the two readings is the temperature rise. Note: The temperatures you measure may be around 160 degrees or more. Be sure the thermometer you use can take that kind of heat.

Now compare the rise you measured with the specs from your manual. If your readings are out of spec you can adjust them by changing the speed of the furnace blower. Raising the speed will lower the rise, lowering the speed will raise the rise. Adjustment is simple if you know what you're doing. Belt-driven blowers are adjusted at the motor pulley. You increase the diameter of the pulley—and thus raise blower speed and lower heat rise—by backing off on a set screw on the pulley, then twisting the pulley clockwise. Then retighten and recheck the rise. Naturally, turn the pulley the other way if you want to raise the heat rise.

Some blowers are driven directly by the motor shaft, so there are no pulleys to adjust. On these you adjust motor speed by changing the wiring at the motor terminals. A diagram should tell you which connections give which speeds. Look for the diagram on the cover of the terminal board, or somewhere on the furnace.

While you're in there poking around with the motor, you might take the time to consider continuous air circulation. With this system, your furnace blower is running all the time, even when the furnace isn't supplying any heat. A waste of electricity? Not according to heating experts. When the fan shuts off, air circulation stops, and air within the home stratifies. Cold air settles to the floors where you can feel it. Hot air rises to the ceiling where you can't. Result is that you need more heat to feel warm.

Continuous air circulation (CAC) solves this. The fan runs full speed when the furnace is heating, but then kicks down to a lower speed when the furnace cuts off. If your furnace is new, it may already have this feature. If so, be sure to use it. (Many furnaces are equipped with switches that let you turn the CAC feature on and off.) If you don't have it, ask about the possibility of having it installed. You can do the job yourself if your fan is belt driven, but direct drive systems are a little trickier to modify.

Conversion involves buying a new two-speed motor and control, and installing them in place of the old motor and control. If you feel up to it, here's how it goes:

Shut off the power to the furnace and remove the panel covering the blower compartment. Check the old motor for amperage, shaft diameter, full-speed power, and frame and mounting arrangement. Also check the direction in which the blower rotates. Now, go to a heating supply dealer and get a two-speed motor that matches the specs on your old motor. Also pick up a single-pole, single-throw fan control designed for two-speed operation.

Remove the old motor but save its pulley. Put the pulley on the new motor and install the new motor on the old mounting plate. Check alignment of the fan and motor pulleys; outer edges of all flanges should line up. Install the belt. Set the new fan control for a top-speed cut in between 85 and 100 degrees, but don't install it yet.

Connect the new motor to the old wires and turn on the power to check for proper rotation. If rotation is reversed, change wiring at motor terminals as indicated on the motor terminal cover.

Shut off the power again and replace the old control with the new one. Turn on the furnace again and let it run a full cycle. It should run with the blower at full speed whenever the furnace temperature rises to the high speed cut in temperature (85-100 degrees) you set on the new control. At all other times the fan should run at slow speed.

That's about all you can tackle on your own, inside the furnace. Now you might call in an expert and have him check over the furnace to see if it's running at top efficiency. He'll use special instruments to check exhaust gases and drafts. And he'll probably analyze the flame pattern. While he's doing this, ask him about underfiring the furnace. What's underfiring?

Chances are your furnace is oversized for the job it has to accomplish. And this isn't good. Furnaces are most efficient when they run almost constantly. On-off operation cuts into that efficiency. So ideally, your furnace should run just about all the time on the average coldest day in your area. Not the coldest day possible, just the average coldest day.

Usually, though, the heating contractor who installs your furnace will pick one to match the coldest day in history. Then just to be sure, he'll move up in size another 5 to 10 percent. Result? You may get efficiency down around 60 percent when a properly-sized furnace would give you 75 percent or so.

That's where underfiring comes in. Your serviceman can effectively "size down" the furnace by changing the nozzle size (oil unit) or orifice (gas burner). There are limits to how far this can go, though. Reducing the flame size too much will result in heat rising up and out of the combustion chamber because the flame is too small to contact the walls of the chamber. Your service technician will be able to advise you on this matter.

At any rate, underfiring is a good thing to keep in mind if you ever need a new furnace, or if you are having a new home built. Be sure to specify to the heating engineer that you don't want a monster oversized furnace, but rather one that is sized to handle your average coldest day. You may run across a day or two when the smaller furnace can't keep your home in the seventies, but you shouldn't feel any discomfort.

Note: If you're having a house built that may later receive additions in living space, be sure to start with enough furnace capacity to cover the added space.

INSTALLING A HUMIDIFIER

There is another way to save on heating costs. During the winter, the relative humidity inside your home may drop as low as five or 10 percent, about twice as dry as the average humidity in the Sahara. As a result, moisture on your skin evaporates into the dry air at a very high rate. Evaporative cooling makes you feel chilly, even though your thermostat may be set at 72 degrees or even higher. Adding moisture to the air via a humidifier will slow the evaporation rate on your body. You'll feel warmer, even with the thermostat set lower than you had it without the humidifier. We all know how humidity can make us feel warm in the summer. Well, it can do the same thing in win-

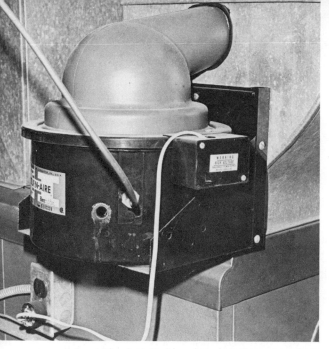

A humidifier installed in a hot air system can reduce heating costs, protect your home and furniture, and even improve your health. This is a return-air model, mounted in the return air plenum. It's wired to run whenever the furnace fan is on.

ter, and save us on heating costs as a result.

There are other bonuses too. Furniture is less prone to shrinkage and cracking. Static electricity is reduced. Your house plants will be happier. And your health may even improve: Colds are more likely to strike breathing passages that have been dried out by low humidity. Since I put in a humidifier, my nosebleeds—caused by the cracking of dried-out nasal passages—have stopped completely.

Before you invest in a humidifier, you'll be smart to check out the humidity inside your home. This will help you decide whether you really need a humidifier, and it may help you pick one with the right capacity.

The only way to measure humidity is with a good hygrometer, and this will cost about $20. Get the kind with a hair element, and stay away from those that cost just a few dollars.

A rule of thumb on selecting a humidifier goes like this: For every room in your house, you need a gallon of moisture a day. But this is a very rough rule. How much moisture a home really needs is determined by the outside temperature, and the moisture tightness of the house. So the rule of thumb needs a little adjusting to make it work.

How tight is your house? If it's a relatively new house, fully insulated, caulked and weatherstripped, you may be able to get by with half a gallon per room. But if the home is old, uninsulated, drafty, and poorly caulked and weatherstripped, you may need two gallons a day per room. Your hygrometer will help you make your decision here, since a very low reading (around 5 percent) indicates a loose home, or one that is very dry for other reasons. A higher reading (10-15 percent) indicates a tighter home, or one that gets a fair amount of humidity from sources such as cooking, bathing, etc.

Some humidifiers are rated in terms of gallons per day, others in terms of pounds per hour. If you want to know how many gallons per day can be produced by a unit rated in pounds per hour, just multiply pounds per hour by 2.9. Your answer will be in gallons per day.

Should you buy a humidifier that

attaches to your furnace, or a self-contained unit? If you have hot water, steam, or electric heat, you have no choice. You'll have to use a self-contained humidifier. Furnace-mounted units will work only with warm-air heating systems. Don't feel cheated, though. A good self-contained, whole-house humidifier has some advantages over furnace mounts. For one thing, a furnace-mounted unit can supply moisture only when your furnace is running, unless you have continuous air circulation. If your furnace is oversized, it may not run long enough during the day to give your humidifier a chance to do its job.

But a whole-house humidifier — not a little room-sized console — independent of your heating system can supply moisture 24 hours a day if required. And humidity has a way of spreading evenly throughout your home without the aid of ducts. So hooking into the heating system has no real advantage.

If you decide against a furnace mount, you can put your humidifier almost anywhere. You can put it in a closet or utility room, and run a short duct for the moisture discharge. Terminate the duct with a floor or wall-mounted register.

Whichever type you decide on, be sure to get one with a humidistat. This is a sort of "moisture-sensitive thermostat." It turns the humidifier on when moisture is needed, and shuts it off when the proper level has been reached. This liberates you from controlling things manually, and

protects against the possibility of overhumidification, which can be even worse than the original arid condition you've set out to correct.

Moisture Vapor Barriers. Before you hook up a humidifier, be sure you have some sort of vapor barrier in your home. Vapor barriers are discussed in the chapter on insulation, and they're vitally important if you decide to humidify. For two reasons:

1. Without one, moisture will probably dissipate out of your home as fast as you can produce it. The money you spent for the humidifier will be wasted.

2. Even worse, as the humidity seeps through your walls, it will eventually reach a cold surface (most likely the sheathing or siding) and condense there. This in turn can cause all kinds of problems including rot, and peeling and blistering paint when warmer weather rolls around. You could find yourself needing to repaint your house every couple years.

INSULATING DUCTS

Any time your furnace ducts pass through an unheated portion of your home, they waste heat. Insulation around the ducting can reduce this waste, and the savings will pay for the cost of the insulation in a short time. If you happen to have central airconditioning, the insulation will pay off even faster.

To do the job yourself, get insulation made specifically for ducts. This comes in one and two-inch thickness, but if you're going to the bother of doing the job at all, you might as well use the thicker stuff. If your ducts are just for heat (not airconditioning) you won't need a vapor barrier. You will need it if you're insulating airconditioning ducts. It goes on the outside. Use tape to seal the seams between blankets; this prevents condensation on the ducts. Don't apply the tape too tightly or you'll compress the insulation and decrease its effectiveness.

If you have hot water heat you can insulate the pipes wherever they run through unheated space. Wrap-on fiberglass, with an overwrap of tape made specifically for this purpose, makes the job fast and easy.

Providing Combustion Air.   Whenever your furnace is running, it needs a steady supply of air to breathe. Where does this air come from? Most likely your furnace sucks it up from all parts of the house. This means the furnace is drawing off air you've already paid to heat. This air is then replaced by cold outside air filtering in through every little crack in your house. Obviously you'll have to burn more fuel to heat this cold air, and this in turn will draw in more cold air.

That's not all. The cold air being sucked into the house creates cold drafts. And heating this air will dry it out, lowering the humidity in your home.

There's a simple solution to this problem. Let your furnace breathe fresh air from outside, and keep your warm air to yourself. You feed your furnace fresh air by running a duct from outside to a location near the furnace. A simple way to do this is to replace a pane of glass in one of your basement windows with a sheet of 1/4-inch plywood or hardboard. Install an ordinary clothes dryer vent in this panel, and run flexible dryer vent hose over to the furnace. If the vent you use has a one-way flapper valve in it, remove the flapper so air can be drawn into the vent. Now your furnace will have a ready supply of outside air, and it won't have to rob you of air you've already paid to heat. Users of this technique report fuel savings of 14 percent and more.

A few cautions: Don't run the fresh air duct directly to the furnace intake. It's better to terminate the duct so that its discharge passes over the furnace smoke pipe. This will warm the incoming air to some degree. Feeding the air directly to the furnace may cool the exhaust gases so much that they condense in the chimney, possibly corroding metal pipes and attacking the mortar in unlined stacks. Also, feeding the air directly to the burner creates another problem. A burner is adjusted for proper air to fuel ratio, but this adjustment must be varied to suit the temperature of the combustion air. A direct line will feed in air at whatever the outside temperature is, and outside temperatures vary so widely that proper fuel-air ratios will be impossible to achieve.

There's another trick you can use to

get a little free heat—and humidity— during the winter, if you own an electric clothes dryer. Dryer makers insist that proper dryer installation always includes venting the dryer exhaust to the outdoors. And much of the time, getting rid of that heat and moisture-laden air is a good idea. But during winter, chances are you can use all the heat and humidity you can get.

Well, venting your dryer inside your home can fill some of that need. Caution: I'm not talking about gas dryers here, only electrics. Venting a gas dryer into the home can fill your home with harmful gases.

There are two ways to go about indoor venting. A simple approach is to simply slip the dryer vent hose free of its connection with the vent mounted where the passage exits your home. Use a large hose clamp, in a size made for dryer hose, to fasten a cloth bag or nylon stocking over the end of the dryer hose. This will serve as a filter to trap lint that would otherwise be blown around your home. Whenever you want to, you can always remove this filter, slip the hose back on the vent, and go back to venting the dryer outdoors. You'll want to do this during warm weather of course. But you may also want to go back to outdoor venting even during the winter if you find your house is getting more humidity than it needs.

A classier approach, and one that allows you to change from indoor to outdoor venting with less effort, is to run a Y-shaped configuration of hoses out of your dryer. Run one arm of the Y to an outside vent, the other to an inside vent. Valves in each of the arms will then allow you to control the path of the exhaust at will.

Note: Dryer makers say that inside venting will slow drying times by forcing the dryer to breathe damp air. If you find your drying rate slowing down, you can run your indoor hose farther from the dryer. This will allow the moisture to dissipate somewhat before it gets back to the dryer's intake. And since the humidity will always be fairly high right at the outlet of the hose, you may want to extend the hose into a more remote part of your home (basement for example) if your dryer is located in the kitchen or other high-traffic area.

RECLAIMING WASTE HEAT

No doubt about it, a lot of the heat produced by your furnace goes right up the flue, on into the open air where it's wasted heating the great outdoors. Actually, not all this heat is wasted. Some of it is required to create the draft that feeds fresh air to your burner's combustion chamber. Nevertheless, you can reclaim some of this heat by adding attachments to the metal smoke pipe exiting your burner. The simplest device is a series of metal fins—sort of like a ruffled clown's collar—that you can slip around the pipe to increase its surface area. You can increase the effectiveness of this setup by using a small fan to boost the airflow over the finned surface. You can do this also by special fan-boosted heat exchangers designed to hook right into the flue pipe. Some of these have finned exchanger surfaces; others use a heat pipe principle, carrying the heat out of the flue via evaporation and condensation of a liquid inside enclosed tubes.

The only real problems with these devices is that they can do too good a job. They'll take off so much heat that the draft is weakened. Windy days may result in downdraft conditions which in turn mean incomplete combustion, and the possibility of fumes in your home. In extreme situations the overextraction of heat can result in the condensation of flue gases inside the chimney. Corrosive acids form and can eat away at flue pipes.

These problems are most severe with modern efficient furnaces that have little waste heat going out the chimney. Older furnaces may have so much waste heat that reclaiming can't cause a problem. Fortunately this works out to an ideal situation; the furnaces that can benefit the most from heat reclaiming are also the most tolerant of it. Before you go to the bother and expense of heat reclaiming, have your furnace checked by a heating technician. You should do this anyway. As described earlier in this chapter, he'll check draft and flue temperature to determine your furnace efficiency, then perform a tuneup if required. While he's doing this he can also advise on the wisdom of installing a heat reclaiming device.

Even after your furnace shuts off, heat still continues to rise out the chimney. This heat, coming from the hot walls of the combustion chamber, is 100 percent wasted. There are several automatic dampers designed to prevent this loss. When the furnace shuts down, the damper closes off the flue, preventing the loss of heat. It also prevents the loss of heated room air. Test reports on very old furnaces indicate that savings of up to 40 percent. But modern furnaces will show much less improvement, and flue dampers have serious drawbacks.

The safety problem. Any time you have combustion going on in the furnace while the flue damper is closed, you're in for serious trouble. With the flue closed off, the exhaust gases—including carbon monoxide—have nowhere to go but into your home.

Of course flue dampers have safety interlocks to prevent firing of the furnace if the damper is closed. But even if these interlocks are redundant and fail-safe, problems can still occur because the furnace itself doesn't have the necessary safety features. Under certain circumstances, oil can trickle into the combustion chamber even when the furnace is off. The same type of thing can happen in a gas furnace if a valve fails. The safety interlocks on the flue damper can do nothing in such a situation. The solution is a whole new system, with both the furnace and the flue damper designed to work together, so both are equipped with interlocks. Furnace systems of this type are currently under development.

As it stands now, you're better off forgetting about flue dampers until the complete systems described above are on the market.

UPGRADING YOUR FURNACE

If your heating plant is very old, updating may do a lot to increase its efficiency. In my area of New England there are a lot of old coal burners, converted many years back to burn oil. The old oil burners that were installed back then are not as efficient as many of today's designs. Replacement with a modern unit would save money and oil in the long run. A fuel oil dealer in my town is even offering a deal on a new high-efficiency burner in order to attract new customers. A cut-price offer of this type makes the idea of conversion even more attractive.

A new type of oil burner nozzle that may reach the market soon promises to be even more efficient and trouble-free than the best types available at present. This Babington superspray nozzle could give an estimated 15-percent savings in oil, and it will be adaptable to most oil burner heads being used today.

Meanwhile, for homes equipped with gas-fired furnaces there's the revolutionary heat-transfer-module furnace from Amana. The whole furnace is about the size of a basketball, and it operates at about 82-percent efficiency. This is far greater than the 60 to 70 percent efficiency of most furnaces in operation today. Drawbacks? There are two. First, this is a whole new furnace, not a new burner to update an old furnace. Second, it is available only as part of a combination heating-airconditioning package. To get it you have to scrap your old furnace, and install central airconditioning.

HOME COOLING

In a recent four-year period, the number of U.S. homes with central airconditioning tripled. For much of the country there are cheaper, more energy-efficient ways of keeping cool. Applied to airconditioned homes, these energy-efficient ways reduce airconditioner operating costs too.

Americans use more electricity to run their airconditioners than is produced in all of China. And while not so many years ago we all got along fine without airconditioning, today most Americans consider it a necessity rather than the luxury it is. Luxury or necessity, however you classify airconditioning, the fact remains that it consumes a lot of electricity. And it's an area of energy use in need of conservation measures.

The airconditioner manufacturers realize this, and they're doing everything they can to make more efficient products. You can do your part too. First of all, don't buy or use an airconditioner unless you really need one. There are cheaper, less energy-consumptive ways of keeping cool, which we'll discuss later in this chapter. By all means try them before you turn to airconditioning. Even if they don't solve your cooling problems entirely, they'll still take some of the load off any airconditioner you decide to use later — saving energy and money all the while.

ENERGY EFFICIENCY RATING

Then, if you decide you really need

an airconditioner, think small. Refrigerating an entire house is wasteful. Small window units can cool one or two rooms to which you can retreat when the rest of the house gets uncomfortable. And while you're thinking small, think "efficient." Efficiency of an airconditioner is expressed by a number called the Electrical Efficiency Rating (EER). The EER is simply the cooling capacity of the unit in Btu's divided by the unit's wattage. All airconditioners sold today must have the EER labeled on the machine. But if you buy a used unit with no EER label, remember that you can compute it yourself by dividing Btu's by watts.

What's a good EER? Well, it varies with unit size. Tiny portables might be considered good at an EER around seven, but a large room unit would have to rate over 10 before it would be considered highly efficient. Central units peak around 10 or 11.

If you want to figure how much it will cost you per year to run an airconditioner, multiply the unit's wattage rating, by the hours the unit runs (1,000 is about average), by your cost for electricity in dollars per kilowatt hour. Divide this number by 1,000 and you have your yearly operating costs.

Extra efficiency almost always costs you extra money. A higher EER means a higher initial purchase price. If you want to find out how many years it takes a high-efficiency unit to pay back its extra cost over a cheaper low-EER machine, just compute yearly operating costs for both machines as described in the above paragraph. Then divide the yearly savings of the more efficient unit into the difference in purchase price between the high and low cost machines.

Chances are you'll find the high-efficiency machine pays off in three to eight years in the South, where running time is long. But it may take 20 years for such a machine to pay for itself up north. Since the life span of a typical airconditioner is only 10 to 15 years, the high EER wouldn't pay off. But if you live where it takes this long to make up the difference, you really don't need an airconditioner; you'll only be using the machine about eight days a year. Of course all these calculations are based on the ridiculous assumption that your electrical rates will remain stable. No doubt, however, they will rise, and as they rise the economics of the efficient unit get better and better.

Buying the right size airconditioner is important, and fortunately, it's better to err on the small side than it is to get one that's too large. This is because much of an airconditioner's ability to keep you cool comes from its dehumidifying abilities. A large unit will cycle on and off so often that it never gets a chance to dry the air. A smaller unit will run longer and do a much better job of dehumidifying. While your dealer can help you select the right size for your needs, you can do an accurate job yourself by consulting a copy of the "Cooling Load

Estimate Form," available from the Association of Home Appliance Manufacturers, 20 N. Wacker Drive, Chicago, IL 60606.

TUNING AIRCONDITIONERS

If you have to use an airconditioner, you should always keep it in tune. This is a simple job consisting of a few easy steps, many of which need be performed only once a year. Other steps — such as cleaning filters — should be performed at least monthly.

If you have central airconditioning, shut off all the power to the unit. Check to make sure that the fan belt on the blower is properly tensioned. Press down on the belt halfway between the pulleys and measure the deflection. It should run around 1/2- to 3/4-inch. If it's loose, tighten it. If it's too tight, it can create excessive loads on the bearings. So don't overdo things.

Airconditioning requires greater air flow than heating. If your blower has an adjustable pulley, make sure it's set to run at the highest speed. To make the adjustment, loosen the set screw on the pulley and twist the outer flange so that it is as close as possible to the inner flange. This narrows the pulley and forces the belt to ride higher. Check belt tension after making the adjustment.

Clogged filters are probably the biggest wasters of energy your cooling system has to tolerate. Filters

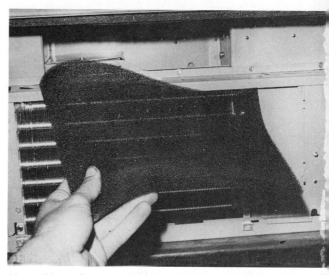

Airconditioner filters should be cleaned or replaced every month. This is a foam type, cleaned by rinsing in water and wringing dry. Clogged filters raise the cost of cooling a home.

Bent evaporator fins cut down on air flow and cooling efficiency. When just a few fins are bent, you can straighten them by careful picking with a pencil or piece of wood. Don't use metal instruments. If large areas of the coil are bent, get a fin comb to do the straightening.

should be cleaned or replaced once a month, but most homeowners don't do the job at all. While you're in the cleaning mood, clean the condenser and evaporator coils. Use a commercial coil cleaner, available at a refrigeration supply house. While you're there, pick up a fin comb—a simple plastic tool used to straighten bent fins.

To use the cleaner, spray it onto the coils and fins and let it set for about 10 minutes. Then rinse with a hose or a spray of clean water from a squeeze bottle. If the coils are still dirty, give them a follow-up treatment. To use the fin comb, match the teeth-per-inch rating on the comb to the fins per inch on your airconditioner's coil. (The comb will have different sections with varying teeth spacing.) Then run the comb through any bent fins.

Last, check to make sure that air flow over the outdoor condenser coil is not blocked by bushes or other vegetation. Shading the condenser will raise efficiency about two percent, as long as the air flow is not blocked. Small evergreens planted a safe distance away from the condenser will provide the shade. Just be sure to prune them back if they begin to choke off air to the coil.

Room airconditioners should get the same treatment as central units. Clean the filters and coils, oil the motor if it has oil fittings. Straighten bent fins, and check outdoors to be sure the air flow through the unit is not being impeded.

Adjusting Air Flow. If you have a window unit, and if it has adjustable louvers to direct the air flow, set them to force the cool air upwards. Cool air naturally drops to the floor. Aiming it upwards will help distribute the air before it has a chance to stagnate at your feet. Unimpeded air flow is as important inside the house as it is outside. Be sure no furniture or curtains block the flow of the cool air.

Central cooling units can be adjusted to distribute cool air just where you need it. In our discussion of heating efficiency we described how to adjust dampers in the ducts to provide warm air to those rooms that needed it most. You can do the same thing with central airconditioning. Rooms requiring one warm-air setting in winter will require a different cool-air setting in summer. So you'll have to change the damper settings every time you switch from heating to cooling. That's why it's a good idea to mark the position of the damper handle once you've found its ideal setting for heating. Then mark it again at its best setting for cooling.

Again, it's best to close down on the dampers in ducts leading to rooms that need little cooling. This will provide more cool air to the rooms that need it most. The general rule is that rooms where there is a lot of activity or heat (kitchen, for example) will need the most cooling. Other rooms may need less cooling.

Most air registers are located near the floor. This is fine for heating, but not so good for cooling. The cool air comes out at floor level and stays there. It's picked up by the return air duct and rerouted through the cooling system again. You keep cooling the same air over and over, and it never gets high enough in the room to have any cooling effect on most of your body. You can remedy this situation to some degree by installing clip-on or magnetic baffles over the registers to direct the air flow upwards. Of course you'll want to take these off during the heating season or you'll wind up with all the hot air hugging the ceiling.

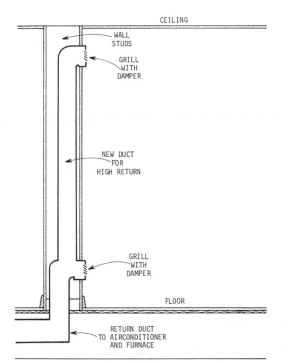

A high-air return gives better distribution of cooled air with central airconditioning. During the summer, open the top grill and close the bottom grill. During the heating season do the opposite, opening the bottom return and closing off the top.

An even more effective way of spreading cool air throughout all levels of a room is to use a high-return air duct system. In this system, a return air duct is placed up high—near the ceiling—where it draws off the hottest air in the room and sends it to the airconditioner for cooling. To install a high return you remove your finished wall surface between the two studs that straddle a low-return air duct. You then install a sheet metal liner or rectangular duct leading from the low duct up to a position near the ceiling. Install a new register grill at the high return and cover the wall again. Now you can switch between the high return for cooling and the low return for heating by closing off one or the other with a metal plate. Or you can install grills with built-in dampers to achieve the same effect.

ALTERNATIVES TO AIRCONDITIONERS

Many other techniques that will increase cooling efficiency have been covered already in the chapters on insulation and heating. Included are adequate weatherstripping and insulation, duct insulation, and caulking. The chapter on windows covers such ideas as double glazing, application of light control film, and installation of storm windows.

Ventilation. But there are some other ideas for cooling that may well eliminate the need for airconditioning. Probably the most important is proper attic ventilation. During the summer, temperatures as high as 130 degrees are not at all uncommon in an attic. This heat in effect turns your ceiling into a radiant heater. With proper ventilation—either natural or fan-induced—you can drop attic temperatures way down. Most effective are fans installed near the peak of your roof, and controlled by thermostat. The 'stat will turn the fan on at a temperature usually around 110 degrees, then cut it off when the temperature drops to about 95.

Roof-mounted vents are more effective than gable-mounted fans because you can position them near the midpoint of the roof's length. Most are sized to fit right between the rafters so installation is quite simple. You cut a hole through the roof, place the unit in position, slipping the attached flashing up under the shingles. Inside the attic you wire the fan to a junction box. For best results, the fan should be teamed up with soffit vents. Installing gable fans is usually even simpler; there's no need to cut through the roof.

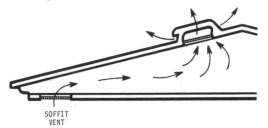

SOFFIT VENT

Powered roof vents with thermostatic controls are most effective when used in conjunction with soffit vents as shown here. A powered vent can lower attic temperatures 30 degrees or more, reduce cooling loads on your home, and prolong roof life. They're easy to install. Remove the shingles from between two rafters. Sabersaw a hold for the vent. Slide the vent into place under the shingles and nail it in place. Inside the attic you wire the fan to a junction box.

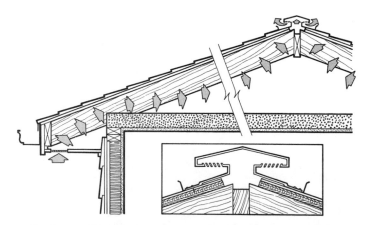

Continuous ridge vents can drop attic temperatures as much as 15 degrees. Total open area of ridge vents should be matched by an equal area of soffit vents. Together they should provide an open area equal to at least 1/300 the area of your attic floor, in the northern half of the US, or 1/150 your attic floor area in the South.

Natural attic ventilation can be increased by installing ridge vents or turbine ventilators. Ridge vents provide a continuous—though weatherproof—opening for the full length of the roof's peak. The total free or open area of the roof vent should then be matched by an equal area of soffit venting to providing a free flow of cooling air. The total area of roof and soffit vents should be equal to at least 1/300 of the attic floor area. Twice that much vent area is even better. In addition to carrying off hot air, vents also exhaust humidity, another cooling bonus.

Before airconditioning came along ventilation was the only way to cool a home. You can set up ventilating air currents with a fan, or rely on natural powers of convection. The trick is to bring in cool air near or below ground level and exhaust it from the attic.

Convection or stack cooling relies on the hot air within your home rising and creating a chimney or stack effect. For best results, open your basement windows to let in cool air. Air is coolest near the ground and near basements. The coolest part of any house will often cool the air even more. Once the cool air is in the house, it should have an unrestricted path in which to travel. Open the door at the head of the basement steps. Make sure all interior doors are open, and open the door to the attic. If your attic has no door—just a hatch or a folding stairway that will inconveniently block up a hallway—put a screened hatch in your upstairs ceil-

ing. Last, provide some way for hot air in the attic to escape. Roof vents, gable vents, ridge vents, or a turbine ventilator will all help get the air moving.

If you decide to use fans or powered vents, it still pays to set your home up as described. The stack effect will relieve the fan of some of its work and help encourage better air flow. Remember, ventilation can help cool your home by getting rid of trapped heat, but it will not make your home any cooler than the temperature outside. On the other hand, the slight breeze created by ventilation will make your home *feel* slightly cooler than the thermometer might indicate.

Roof Color. Everyone knows that a white roof is cooler than a dark one. It reflects a substantial part of the sun's energy. But did you know you can change the color of roofing without reshingling? Roofing paints are the secret. They can go on over shingles, roll roofing, even metal, and they can add years to the life of the roof, both by virtue of protective coating and reflective cooling. Most large paint stores carry roofing paints. So do Sears and Montgomery Ward.

Radiator Coolers. If you happen to have a well, spring or other source of cool water on your property, you can cool your home very effectively without running an airconditioner. One fellow cools his six-room Oklahoma house by pumping well water through three automobile radiators housed in a wooden box outside one of his windows. A squirrel-cage fan

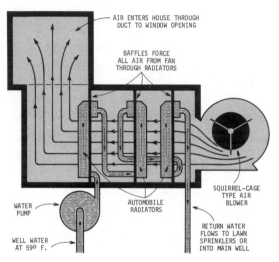

AIR ENTERS HOUSE THROUGH DUCT TO WINDOW OPENING

BAFFLES FORCE ALL AIR FROM FAN THROUGH RADIATORS

SQUIRREL-CAGE TYPE AIR BLOWER

WATER PUMP

AUTOMOBILE RADIATORS

WELL WATER AT 59° F.

RETURN WATER FLOWS TO LAWN SPRINKLERS OR INTO MAIN WELL

Well-water cooling is possible with a setup like this. A 500-gallon-per-hour pump circulates well water through three junked car radiators housed in a plywood box outside a window. A fan blows air through the radiators and into the house through a duct leading to the window. Used water can be used to water lawn and garden. Or it can be returned to the earth via a second well.

forces air through the radiators and into the window. Even in 100-degree weather this system keeps his home down to 70 degrees. Water at 59 degrees is pumped through the radiators at a rate of about 500 gallons per hour. Used water is returned to the ground via a second well.

Roof Sprinklers. Water can cool your home another way. If you fasten a simple sprinkler-type hose—upside down—to the peak of your roof you can use it to trickle water down the roof and set up an evaporative cooling cycle. Connect the sprinkler to an outdoor bib and adjust the flow to use as little water as possible. Just enough to keep the roof wet down to the eaves is best.

More Cooling Ideas. Other ideas for cooling without airconditioning will be presented in the section on energy-efficient architecture in the next chapter.

ARCHITECTURE

If you take the word of energy experts seriously, the time is coming when our conventional sources of energy will either be gone, or priced out of reach. Just how soon this time will come is open to debate, but it could easily come within your lifetime. How livable will your home be when you can't get or afford the fuel to heat it? Or the electricity to light it.

No doubt, the home you live in right now is not designed to cope with the future. And the best way to get a home that can cope with the future is to design and build it from scratch. The basic principle behind the design will be simple: The house must gather as much energy as possible from its environment, and it must then make use of every bit of energy it has gathered.

Helping you design this home is what this chapter is all about. For optimum efficiency, the home of the future should be designed and built from scratch, but you may elect to redesign your existing home. If so, many, but not all, of the following ideas may be of help to you.

Think Small. This is the basic rule for energy efficiency. The smaller your home the less energy it takes to run it. It's that simple. The trick is making the small house livable. Efficient use of space is the key. Take a bedroom as an example. In most homes today, the bed takes up most of the bedroom floor space. The space above and below the bed is wasted. A more efficient design would raise the bed up near the ceiling in a sleeping loft arrangement. Now you have room under the bed for a desk or storage in the form of closets or a dresser. If you must have your bed at conventional height, make it a platform bed with storage drawers underneath. This eliminates the need for bedroom dressers and the space they require.

A separate dining room is a waste of space. You'll be smart to combine it with another room, either the kitchen or living room. If you do this you'll find you can get a comfortable living room/dining room, or eat-in kitchen that requires only two-thirds the space you'd have to invest in two separate rooms. Combining rooms also gives a more open, spacious feel to your home than you'd get chopping it up into separate rooms.

Built-in furniture in any room is another good way to cut down on space requirements. A foam-slab sofa with storage under the seat is one good idea. So is storage built into the knee walls in attic rooms. A well-designed closet can store twice as much as today's typical single pole design.

Use Efficient Shapes. Closely akin to keeping a house small is the principle of keeping its surface area small. Your goal is to enclose as much space as possible with the least amount of wall and roof exposed to the weather. The ideal shape is the dome. It encloses the greatest amount of space with the least amount of skin. Its interior, however, is not as easy to use as rectilinear shapes. A cube is a good second choice, rectangles with low length-to-width ratios are good third choices.

It makes more sense to increase size by adding a second or third floor than it does to expand the floor plan. Homes lose a great deal of heat through the roof. Adding a second floor can double your living space without any increase in roof area.

Building up rather than out also has

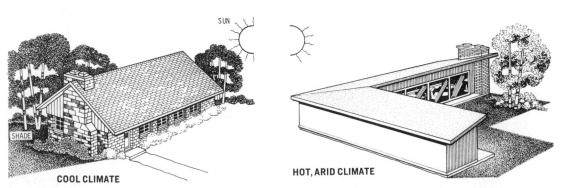

COOL CLIMATE

HOT, ARID CLIMATE

In a cool climate, your home should have as little surface area to lose heat as possible. But in warm climates a large surface area, shaded from sun, will help keep your home cool. No building configuration works well in all climates.

TREE SHIELDS (Overhead View)

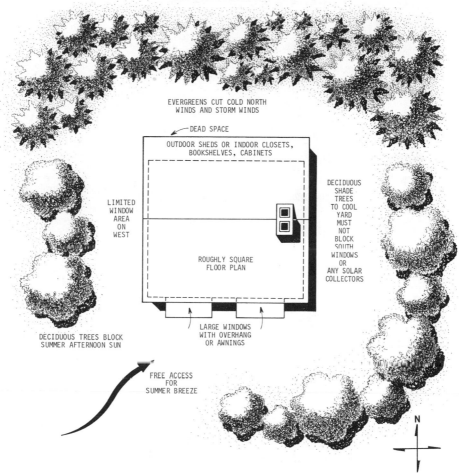

EVERGREENS CUT COLD NORTH
WINDS AND STORM WINDS

DEAD SPACE

OUTDOOR SHEDS OR INDOOR CLOSETS,
BOOKSHELVES, CABINETS

LIMITED
WINDOW
AREA
ON
WEST

DECIDUOUS
SHADE
TREES
TO COOL
YARD
MUST
NOT
BLOCK
SOUTH
WINDOWS
OR
ANY SOLAR
COLLECTORS

ROUGHLY SQUARE
FLOOR PLAN

LARGE WINDOWS
WITH OVERHANG
OR AWNINGS

DECIDUOUS TREES BLOCK
SUMMER AFTERNOON SUN

FREE ACCESS
FOR
SUMMER BREEZE

N

Good site planning takes advantage of beneficial climatic factors and minimizes detrimental factors.

other advantages. If you plan to use solar heat it may put your roof high enough to clear trees or other obstructions to a clear shot at the sun. It also allows you to make good use of convection currents within your home for good distribution of air and for natural cooling in summer. A tall home acts like a chimney.

Orient Wisely. The way your home sits on its land should be determined by the sun and the wind. The main window area of a home should face south. This will let you collect solar heat with your windows. The north side of your home should have no windows at all if possible. In fact, it's best to "sheathe" the whole north side of your

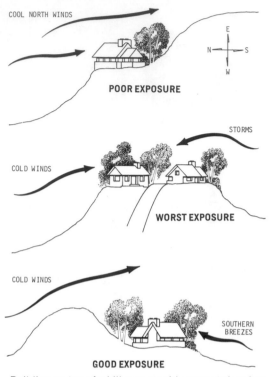

Building on top of a hill may provide a great view, but hilltops offer worst possible exposure during the winter. It's a better bet to place your home halfway down a southern slope.

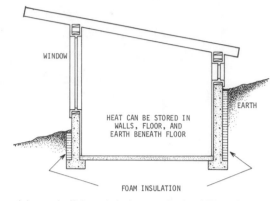

A house built largely below grade should have insulation around the perimeter, but not directly beneath the floor. Earth below the floor can act as heat storage if left uninsulated.

home with unheated storage space. This can take the form of outdoor sheds built against the house, unheated garage space, or at least indoor closets and built-in storage. The object is to create an insulating pocket of dead air all along the north wall.

If you plan rooftop solar collectors, be sure your roof ridge runs east and west. And make the slope of your roof approximately equal to your geographical latitude.

Go Underground. In most of the country, soil temperatures a few feet below the surface remain about 50 degrees all year long. A home built underground can take advantage of this fact to cut heat losses. A good underground design would consist of a concrete floor, placed over perimeter foam-insulation board, and poured concrete walls, insulated on the outer surface with foam. This home would lose little heat in the winter, and would remain cool in the summer.

Drawbacks: Most people don't like living underground. One way around the problem is to build the home into a south slope. Bury the north wall underground. Leave the south wall above ground or only a few feet below grade at floor level.

Superinsulate. The house of the future should be insulated to at least R-20 in the walls, and R-38 in the roof. Using conventional construction techniques, the average home today won't have room for this much insulation.

The standard 3½-inch wall cavity can accept only R-11 mineral wool. Most attic floors or roofs have only enough room for R-19 insulation (using mineral wool). The answer is to use either plastic foams instead of mineral wool, or to use them together. Best bet for a foam-only insulation job would be to have U-F or urethane foam shot into the 3½-inch wall cavities during construction—after the sheathing and interior finish walls have been placed, but before the siding is nailed on. Or in new construction you can provide more space for insulation by framing the house with 2x6s.

To combine mineral wool and foam, use the new tongue-and-groove foam sheathing system developed by Amspec, plus R-11 mineral wool in the wall cavities. In some cases you'll be able to use the foam sheathing instead of plywood sheathing; so, little extra labor is involved. In other cases the foam will go on over the sheath-

ing. The combined R-values of the foam and wool will come to only about 16 or so, but that's not the whole story. As much as 45 percent of the heat lost through a wall is conducted through the wall studs. Foam sheathing effectively isolates the studs from the outdoors and prevents much of the heat loss. So foam plus wool gives an effective insulation value very close to R-20 when this factor is added in.

Placement is one factor often overlooked when insulation is discussed. Ideally, insulation should go on the outside of your home. Outside placement allows the entire structure inside the insulation barrier to become, in effect, a heat storage mass. In winter, your walls can store heat entering windows. Then whenever the air temperature in the home drops, heat is released from the mass of the home. This not only gives you "free" heat storage, it gives you a temperature buffering system that eliminates, to some degree, wild temperature fluctuations inside your home.

Insulation is not placed outside most homes today because it just isn't convenient to do so, especially with batt and blanket insulation of mineral wool. And no matter what type of insulation you use, it must be protected from the weather. This means some sort of outer skin is called for. But the job can be done. The Everett Barber home in Connecticut is a good example. Its heavy, heat-storing concrete walls are coated on the outside with a three-inch layer of urethane

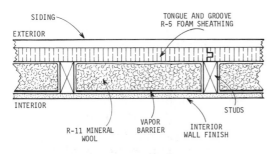

WALL CROSS-SECTION (Top View)

This is cross-sectional view of a wall insulated with mineral wool and foam sheathing. Note that the foam insulates the studs from the outdoors, increasing insulation effectiveness beyond the R-16 (R-11 + R-5) value to about R-20.

foam. R-value is about 18. The urethane must be protected from the sun
to avoid its breakdown. Possible cov-
erings including stucco, T-111 plywood, as well as others.

Use Windows Economically. Glazing can
make or break your home. Well designed, well placed windows will provide heat in winter, cool breezes in
summer, and light all year around. A
bad window can do just the opposite.

Windows designed to collect heat
should face south, or as nearly south
as possible. They, like all windows in
your home, should be double glazed.
To let in the winter sun but exclude
the summer sun, south windows
should be fitted with an overhanging
awning that blocks high-angle summer sun and allows low-angle winter
sun to enter the window. And the

windows should be operable so you
can open them for circulation.

If you plan to make use of extensive
glass areas on your south wall (window area equal to one-tenth of your
floor area or more), you should probably be sure to provide some sort of
heat sink to soak up the excess heat
you'll be collecting during the middle of the day. See the accompanying
illustrations for ideas on making your
own heat sinks.

Due to sun angles, it is much
harder to use shading devices on east
and west windows than it is on southern windows. The sun is just too low
in the east and west to exclude it with
overhanging devices. And overheating from the sun's shining in west
windows can become quite severe on
summer afternoons. Best bets are
trees, fences, or indoor sunscreens
such as blinds, insulating panels, curtains, and shades. Remember, however, it's always best to intercept the
sun outside your windows rather than
inside.

Take Advantage of Landscaping. Using
trees and shrubs for sunscreens is a
cheap, natural way to temper your environment. But landscaping should
not be limited to sun control alone. It
is also an effective way to control
wind. In one study made in Ne-
braska, it was discovered that a windbreak of trees can reduce winter heating costs by as much as 30 percent.

Environmental control with trees
has only one drawback: Sometimes
your overlapping goals of sun and

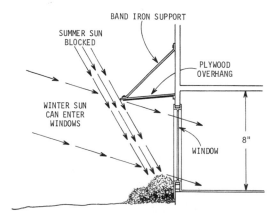

A simple plywood overhang can exclude summer
sun, yet still allow winter sun to enter south windows
for collection. For windows that extend near ground
level, a 4-foot overhang is about right for 40 degrees
north latitude. Stretch this to 4½ feet if you live in
southern Canada, cut it back to 3½ for southern states.

A heavy, stone heat sink can store heat collected through south windows. Then at night the stone will release heat to buffer temperature fluctuations in the home. Field stone, concrete block, or brick can be used for the sink. For the open look, shown here, decorative pierced concrete blocks were used. Dark paint improves absorption of heat. Native stone can be darkened with a masonry sealer, without destroying its natural look.

wind control interfere with one another. A tree planted to provide shade also blocks cooling breezes. Or a windbreak may interfere with solar heat gain. This doesn't happen too often, because windbreaks usually go to the north and west of your home, where they also provide needed shade (west) or can't interfere with the sun (north). And since summer breezes are often south or southwest winds, the trees blocking winter winds rarely interfere with them. A careful study of your building site will help you discover any possible conflicts so you can resolve them before it's too late.

Remember, trees can cool your home even without shading it di-

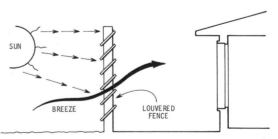

Slatted louver-type fences can provide effective sun screens when space is tight.

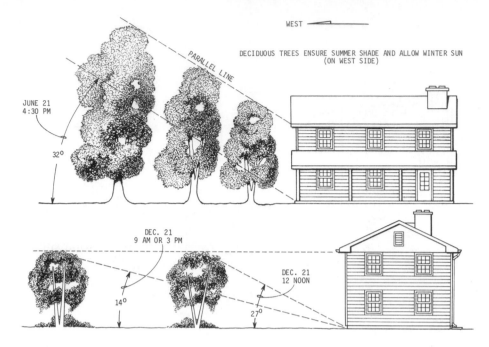

WEST

DECIDUOUS TREES ENSURE SUMMER SHADE AND ALLOW WINTER SUN
(ON WEST SIDE)

PARALLEL LINE

JUNE 21
4:30 PM

32°

DEC. 21
9 AM OR 3 PM

DEC. 21
12 NOON

14°

27°

SOUTH

TO AVOID SHADING WALL (ON SOUTH SIDE)

Sun angles are critical when locating shade trees. This table at left gives sun angles to help plan your plantings. To figure proper planting distances, make scale drawings as shown above, and draw in the sun angles with a protractor.

LATITUDE	DATE	TIME AM	TIME PM	SUN ANGLE	DATE	TIME AM	TIME PM	SUN ANGLE
24°N	June 21	6	6	9.3	Dec 21	7	5	3.2
		7	5	22.3		8	4	14.9
		8	4	35.5		9	3	25.5
		9	3	49.0		10	2	34.3
		10	2	62.6		11	1	40.4
		11	1	76.3		12	12	42.6
		12	12	89.5				
32°N	June 21	6	6	12.2	Dec 21	8	4	10.3
		7	5	24.3		9	3	19.8
		8	4	36.9		10	2	27.6
		9	3	49.6		11	1	32.7
		10	2	62.2		12	12	34.6
		11	1	74.2				
		12	12	81.5				
40°N	June 21	5	7	4.2	Dec 21	8	4	5.5
		6	6	14.8		9	3	14.0
		7	5	26.0		10	2	20.7
		8	4	37.4		11	1	25.0
		9	3	48.8		12	12	26.6
		10	2	59.8				
		11	1	69.2				
		12	12	73.5				
48°N	June 21	5	7	7.9	Dec 21	9	3	8.0
		6	6	17.2		10	2	13.6
		7	5	27.0		11	1	17.3
		8	4	37.1		12	12	18.6
		9	3	46.9				
		10	2	55.8				
		11	1	62.7				
		12	12	65.4				
56°N	June 21	4	8	4.2	Dec 21	9	3	1.9
		5	7	11.4		10	2	6.6
		6	6	19.3		11	1	9.5
		7	5	27.6		12	12	10.6
		8	4	35.9				
		9	3	43.8				
		10	2	50.7				
		11	1	55.6				
		12	12	57.4				

rectly. Their shade prevents the earth from absorbing heat, their transpiration process works like evaporative cooling. A grouping of deciduous trees can lower air temperatures around your home by as much as 15 degrees during the summer. They'll also cut down on noise and dust pollution, making it more pleasant to open your home to the outdoors for natural cooling.

Use the Lay of the Land. If sections of your building site slope, you can often put the slope to work, keeping your home comfortable. When you decide to build on a slope you have two factors to consider: wind and sun.

The wind is a cooling force, so exposure to it will result in a cooler home. For optimum exposure to the wind you should build on a slope facing the wind, or at the top of the slope. This is a good idea only when you live in an area with year-round high temperatures. In most of the country you'll be looking for protection from chilling winter winds, and for exposure to summer breezes. This is most often a south or southwest slope, also an ideal location for other reasons: It gives you good exposure to the sun for solar heating, and it allows you to build at least part of your north wall underground.

The sun is a warming force. Four simple rules describe how different exposures will affect your home. North-facing slopes are cold, receiving little sun. South-facing slopes are warm. Slopes that face east will be

warm in the morning, but will not get much sun after noon. They won't overheat in the late afternoon, but also are not satisfactory for optimum solar heating. West-facing slopes may be cool all morning, then overheat in the afternoon.

All in all, building on a slope facing south makes the most sense for the bulk of our population. This provides well for solar heating and the use of windows to gain solar heat even if true solar heating systems are not used. It avoids exposure to cold winter winds, yet permits summer breezes for free cooling. It avoids late afternoon overheating. And it allows a north wall below grade.

Make the Most of Venting. Before the advent of airconditioning, we all managed to survive without it. We cooled our homes by opening windows and vents, and thereby encouraged air movement. These techniques still work today. In planning your home, provide windows or vents that allow for cross ventilation. Provide small air intake areas on the windward side of your home, larger exhaust areas on the downwind side. The reason for sizing these vents in this way is simple. The small intake area forces incoming air through a constriction. This increases its velocity, and the faster the air moves, the cooler it makes you feel. The large exhaust ports allow the slight vacuum that forms downwind from your home to draw air in through the intakes.

In many cases, intakes will be on

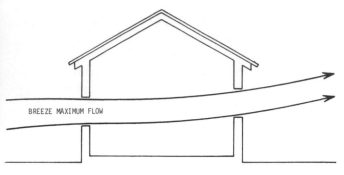

1. INLET-OUTLET AREAS SAME SIZE FOR BEST VOLUME FLOW

BREEZE MAXIMUM FLOW

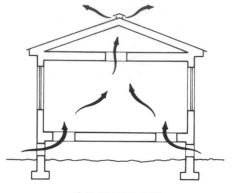

HOUSE WITH CRAWL SPACE

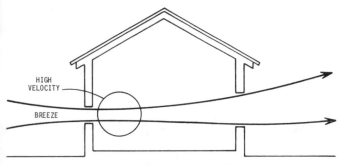

HIGH VELOCITY

BREEZE

2. INLET ONLY HALF OUTLET SIZE FOR MAXIMUM VELOCITY OF FLOW

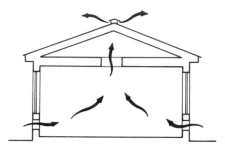

SLAB-FLOOR HOUSE

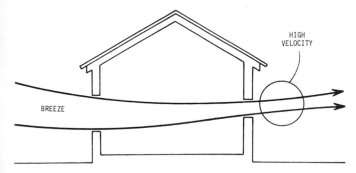

BREEZE

HIGH VELOCITY

3. LARGE INLET/SMALL OUTLET GIVES POOR FLOW

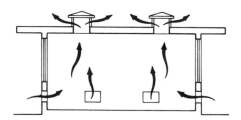

FLAT OR RIB-ROOFED HOUSE

For maximum movement of air through the house by cross ventilation, the inlet and the outlet should be of equal size. When humidity is high, velocity of airflow is more important than volume, so make the outlet larger than the inlet to increase velocity inside the house. Sizing the inlet larger than the outlet reduces both volume and velocity.

Convection can provide the power to move fresh, cool air through your home during the summer. As the illustrations show, you get best results by bringing fresh air in at a very low level, and by discharging hot air through the roof peak.

the south side of your home, exhausts on the north side. Since you don't want to place much window area on the north side, the exhaust ports might better be vents with heavy, insulated doors that can be closed in winter. The exhaust port doors should be the full thickness of your walls, and filled with foam or mineral wool, and weatherstripped, of course. Intake ports might well be small windows.

If you build your home up instead of out, plan to take advantage of stack cooling. Allow air to pass freely from one floor to the next. Provide a cupola or belvedere at the peak of the roof to draw off hot air. Provide air intakes down low, preferably in the basement where the air is coolest.

Be sure to provide venting between the roof and the insulation that lies beneath it. Air movement

through the space between insulation and roof draws off excess heat and moisture. A one-inch air space above the insulation will provide for better cooling than you'd get if you filled the space tightly with insulation. To encourage air flow through this space, provide vents in the soffit—also ex-

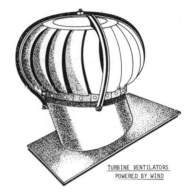

TURBINE VENTILATORS
POWERED BY WIND

Turbine ventilators powered by wind are available in either plastic or metal models.

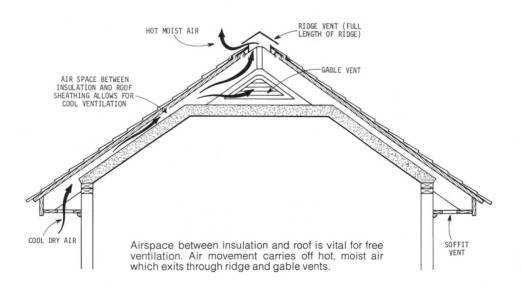

HOT MOIST AIR

RIDGE VENT (FULL LENGTH OF RIDGE)

GABLE VENT

AIR SPACE BETWEEN INSULATION AND ROOF SHEATHING ALLOWS FOR COOL VENTILATION

COOL DRY AIR

SOFFIT VENT

Airspace between insulation and roof is vital for free ventilation. Air movement carries off hot, moist air which exits through ridge and gable vents.

haust vents, either at the peaks of the gables or at the ridge of the roof.

If you have an unheated attic, be sure to ventilate it during the summer months. Wind-powered turbines, thermostatically-controlled attic fans, or ridge-and-soffit venting are three ways to get the job done. The fans are the most effective, but they do require electricity. The other two options do not.

Gather Natural Light. Any time you can light your home without flipping a switch you save energy. Skylights and clerestories are your route to free lighting. Skylights go in the roof. Clerestories are long slim windows that fit in walls, up near the ceiling. Both are positioned up high where they can gather light (which always comes from above). In planning your home, put your light gatherers where they'll provide the light you need for daytime tasks. Kitchens and work areas are prime examples.

To increase the effectiveness of skylights and clerestories, paint the interior of your home in light colors. This will prevent the light from being absorbed before it does you any good. This rule won't limit you to a look of sameness throughout the house. You'll want to use dark colors to absorb solar heat wherever you install solar windows. In these areas, however, you'll have so much glass area, the dark colors won't seem drab, for they'll be flooded with sunlight much of the time.

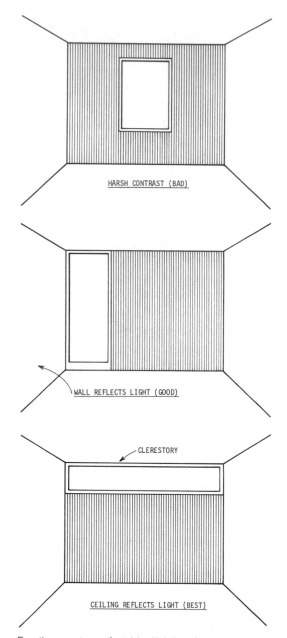

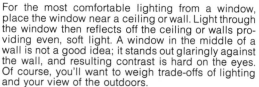

For the most comfortable lighting from a window, place the window near a ceiling or wall. Light through the window then reflects off the ceiling or walls providing even, soft light. A window in the middle of a wall is not a good idea; it stands out glaringly against the wall, and resulting contrast is hard on the eyes. Of course, you'll want to weigh trade-offs of lighting and your view of the outdoors.

Naturally you won't want to lose the energy you saved by gathering free light to escape through the sky-lights and clerestories in the form of heat. So both forms of light gatherers should be double glazed. In addition to the twin glass layers, consider in-stalling foam-lined shutters or other forms of movable insulation that can be closed when the sun goes down. I've triple glazed some of the sky-lights in my home by slipping a piece of clear acrylic plastic into the base of the skylight opening and securing it with a frame of quarter round mold-ing. Preventing heat loss through sky-lights and clerestories is especially important because they are located up high where heat collects that can easily escape if not stopped by good insulation.

Use Efficient Electrical Lighting. Too many homes today have blanket lighting. A high-wattage fixture, or a combination of fixtures, are used to light an entire room. Often this is a waste of electricity. Lighting just a small area of that room will suffice. A better way to go is *task lighting*— providing only the light you need to accomplish the job. For example, you can read a book or work at a desk with lighting from a single 60 watt bulb (even less). No need to light up the whole room. The same principle ap-plies to many other situations.

Limiting the areas you light is only one way to achieve efficient lighting. Another way is to use efficient light sources. Fluorescent lamps are one example. So are the Tensor-type low voltage lamps. A track lighting system offers the flexibility you need for task lighting, and wired to a dimmer switch, it can be adjusted to the low-est suitable brightness for the job and time of day. This cuts consumption, too. Dimmers, however, are not as ef-ficient as they may seem. A 100-watt bulb dimmed to the level of a 60 will use less electricity than a 100-watt bulb at full power, but it will still use more electricity than a 60-watt bulb.

Consider Your Doors. Careful planning can eliminate most of the heat loss as-sociated with doors. Locate main doors out of areas of high or low air pressure. This means keeping them away from windward and leeward surfaces (winter wind). This simple trick minimizes the forces that will force hot air out around the door, or force cool air in. Storm doors also help, but vestibules are even better

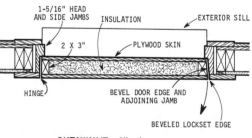

CUTAWAY (Top View)

An extra-thick insulated door must have a beveled lockset edge to provide proper clearance when the door swings. The jamb must be beveled to match the door. The door can be made with 2x3 framing, decora-tive plywood skins, and insulation.

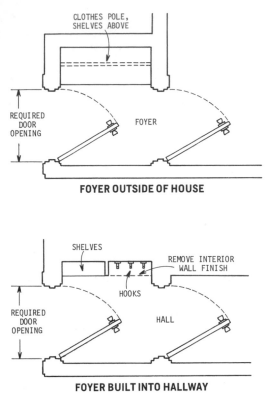

FOYER OUTSIDE OF HOUSE

FOYER BUILT INTO HALLWAY

An entrance foyer can be added onto the outside of a home, or added inside if there's a hallway directly inside the door. Foyers prevent massive heat losses whenever you enter or exit your home. They also provide convenient storage areas.

since they provide an air lock. Even when the outer door is open as you enter or exit, the inner door remains closed.

You might consider building a door that is as thick and well insulated as the wall around it. This will create a few small problems. The door will require a carefully-beveled lockset edge to swing properly. And you'll have to custom-build your own jambs

to accept the door. A simpler expedient would be to foam face the door on both sides and cover the foam with plywood. This will add little weight to the door, so its ease of operation will not be affected.

Consider Adding a Greenhouse. Usually, a year-round greenhouse is more of an energy liability than an asset. But a properly-designed and properly-used greenhouse can be worthwhile.

The greenhouse should be a lean-to type, mounted against the south wall of your home. It should be double glazed to minimize heat loss. It should be connected to the interior of your home via a set of high and low vents. During the winter it should be used to grow only those plants that can stand fairly low temperatures. If you try to maintain a 70 to 80 degree temperature during the winter, the greenhouse will cost you energy. The rule on growing only hardy plants is not much of a limitation since you can grow a variety of cool-weather crops such as lettuce cole crops, and many root crops. Overall, this greenhouse may lose more energy than it collects, but the value of the crops should offset the loss.

Size Your Physical Equipment Properly. In the chapter on heating efficiency, I pointed out the importance of sizing your heating plant to match its load. Oversizing costs you money two ways: You pay more for a larger furnace than you would otherwise need, and then you have to live with inefficiency and fuel bills that are higher than they should be. So make absolutely certain not to oversize your furnace when you build a new home. If necessary, call in a heating engineer and have him run a heat loss calculation to determine the best size for your furnace. And don't forget to tell him not to oversize. Let him know what you're doing. Tell him you'll be happier with something a little undersized than with too big a burner.

This principle is especially important if you're going to use the furnace only as supplementary heating to go with a solar heating system, or wood heat of some kind. If, for example, the furnace is used mostly to keep the home from freezing up while you're not at home to feed a woodstove, it will make little difference if the furnace is too small to hold the house at 72 degrees. It will still prevent freezing, and you won't be home so you won't care if the house is chilly.

Don't Forget Hot Water. Some 15 percent of the energy used in your home goes for heating water. That makes water heating an obvious place for energy savings. The biggest savings are possible with solar water heating. But there are other ways to save. One is to build a tempering tank into your hot water system. This will bring water temperature up to room temperature before it enters the water heater. As a result, the load on the heater is decreased. In the summer,

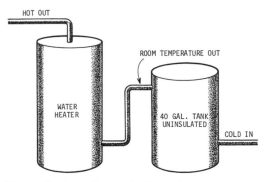

A tempering tank allows water to reach room temperature before it reaches the water heater. This takes some of the load off the water heater. Be sure the water supply enters the tempering tank at the bottom and that the outlet is near the top. In turn, water leaving the top of the tempering tank should enter the water heater at the bottom.

the water in the tempering tank is warmed for free by the naturally warm air in your basement. In winter the warm air in your basement comes by courtesy of your heating system, so the water is now being tempered by whatever fuel you use to heat your home. Thus, winter tempering isn't free, but it also doesn't cost you any more than you'd have to pay to heat the water entirely with your water heater. In fact, if your water heater is electric, you'll probably find that tempering the water with oil- or gas-heated air will still save you money over the cost of electrically heating the water.

Other simple ways to save on water heating involve the various devices that cut hot-water consumption. Fine-spray shower heads, aerator nozzles, and foot-controlled faucets for the sinks are all very easy to install in a new home.

PART TWO

SOLAR ENERGY

Every fifteen minutes, enough solar energy reaches the earth to supply the entire world's energy needs for a year. Many enlightened people feel we ought to collect that energy and put it to use.

Solar energy is free. It's plentiful. It doesn't pollute. So why don't we put it to better use? Two reasons: First, it comes to us in diffuse form. Sure there's a lot of it, but it's all spread out. This means that to collect enough of it to do useful work, we have to gather it over a comparatively wide area. And that costs money. Let's take a look at an example. To heat your home by the sun, you need sunlight collectors that cover the equivalent of roughly half your home's floor area. If you have a 1,000-square-foot home, you'll have to collect the sunlight falling on an area of 500 square feet. The average costs for today's solar collectors run about $10 per square foot. This puts your collector costs for a 1,000-square-foot home right around $5,000.

The second reason for the rather limited development of solar energy? It's an intermittent energy source. When the sun goes down at night, or spends the day behind a thick blanket of clouds, it provides us with very little energy. We solve the problem—to a degree—by storing energy collected during the sunny hours. Then we have it whenever we need it. But again, storing energy costs money.

Up to now, it has simply been cheaper to burn fuel in a $1,000 furnace than it has been to collect free sunlight with a $6,000 solar heating system. But the picture is changing now. Every rise in fossil-fuel prices makes solar energy look more attractive and practical.

Space and water heating by solar means has now become economically practical over much of the U.S. And where it might not be practical today, it probably will be in years to come. Sooner or later, petroleum will price itself out of the home heating picture.

Various respected estimates terminate the world reserves of petroleum prior to the year 2025. That means it is important to begin the changeover to solar energy as rapidly as possible. Because solar energy is relatively expensive to collect and store, it is most practical in homes that are built to conserve energy. In the past, very few homes of this type were built. Ideally, nearly every new home built should be equipped with solar space and water heating equipment. Failing that, new homes should at least be built with an eye towards future installation of solar equipment. And energy-efficient architecture should be mandatory.

This won't happen, however. Today's way of looking at new housing won't allow it. Today, the prime attention is focused on initial cost. The object is to make the home cheap to buy. Don't worry about how much it costs you to maintain the home after you buy it.

A more intelligent approach to housing is to look at the life cost of the house. How much will the house cost you to buy *and maintain* over the life of the house? Wouldn't it make good sense to add solar hardware to a home at a cost of say $6,000, if, over the life of the house, the solar hardware managed to save you $10,000 in heating costs? And wouldn't that solar hardware make even more sense, if after 15 or 20 years it were the only way you could heat the house because gas and oil were unavailable?

And wouldn't that solar hardware have even more appeal when you considered it was saving fossil fuels near exhaustion?

One other thought: When the time comes to sell your house, that $6,000 solar heating system has real value. It adds to the value of your house. *You get the money back when you sell.* You sure can't say the same for the money you spend on fuel.

THE SWING TO THE SUN

It would benefit this country and its people if the government would encourage the swing to solar energy. The government could provide low-interest loans for the installation of solar heating equipment. It could exempt solar heating equipment from property taxes. Or the money from income taxes could be spent on solar equipment, and thereby subsidize the solar industry. The government could even *require* solar hardware in new construction.

But something needs to be done soon. The longer we go on building homes the old way, the harder the inevitable conversion to solar energy. As mentioned before, solar energy is most practical in homes designed and built to save energy. Older homes *can* be improved and retrofitted with solar heating equipment. But not without certain sacrifices in efficiency. For example, you can add only 3½ inches of insulation to a wall built on 2x4 studs. There's no room for any more than that. A home built in the wrong place on its lot (with poor solar exposure, or facing the wrong direction) will be a poor candidate for a solar conversion. So will a home with lots of glass on its northern wall, or with a large surface area in relation to enclosed space. Or with a low-pitch roof that won't support solar collectors at the ideal angle.

The point is this: If the average life of an American home is 40 years, every new home we build from now on will become obsolete, totally obsolete within its lifetime, unless it is a solar home.

THE BASIC SOLAR HEATING SYSTEM

A solar heating system is actually a system of four basic subsystems: 1) the collectors, 2) the storage system, 3) the distribution system, and 4) the backup heating system. Let's take a look at each of these components separately.

The Collector. This is the heart of any solar heating system. Its purpose is to collect solar energy, transform it into useful heat, and allow that heat to be drawn off for storage. The most common type of collector for space heating—and for water heating—is the flat-plate collector. At its simplest, the collector is a black plate that absorbs the sun's rays. As the rays are absorbed, they are transformed into heat. Place a piece of black metal outdoors on a sunny day and it will soon be too hot to touch. That's the principle behind the flat-plate collector.

Okay, we've produced the heat we want. Now to draw it off for use. To do this, we circulate a fluid through the collector to pick up the heat. This fluid can be a liquid such as water, or it can be air. If we use water, it can pick up the heat by trickling over the surface of the plate, or by running through pipes built onto or into the plate. If the heat transfer medium is air, it can be ducted over or under the plate, or through it if the plate is perforated. Once the heat is picked up it is carried off to storage.

As soon as the collector plate warms up it will begin to lose heat back to the cool air around it. To prevent this, we put the plate inside a shallow, topless box. We insulate the back of the plate to reduce heat loss in that direction. We place a sheet of glass or other clear material a short distance above the plate to slow heat loss in that direction. Result: increased efficiency.

This brings up an interesting point. You might think that the higher the

LIQUID COLLECTORS

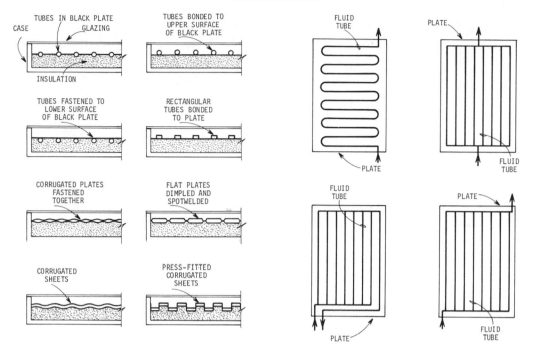

Flat plate collectors can take many forms as these sections and plan views show. Some of these designs can be made in the home shop. Others require factory equipment.

AIR COLLECTORS

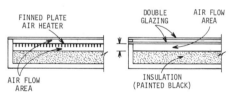

temperature at the collector, the better. Actually the opposite is true. The hotter a collector gets, the more heat it will lose to its surroundings, and the less efficient it will be. So to keep efficiencies high we want the heat-transfer fluid to pick up the heat as soon as it is collected, before it has a chance to build up and leak away.

One more note on collector ef-

ficiency. Efficiency in itself is not an important consideration. What you really want in a collector is a means of collecting large amounts of energy at low cost. Therefore, a collector design that's only 10 percent efficient, but costs only $2 a square foot is a better buy than one that is 60 percent efficient, but costs $20 a foot.

Efficiency becomes important only

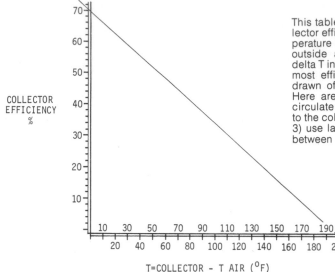

This table shows the direct relationship between collector efficiency and delta T (difference between temperature inside the collector and the temperature of outside air). With this particular collector, cutting delta T in half doubles efficiency. A collector is always most efficient when heat is collected and quickly drawn off before it can build up and radiate away. Here are ways to assure quick pickup: 1) always circulate the coldest air or water from storage back to the collectors for reheating; 2) keep flow rates high; 3) use large heat exchangers; 4) assure good bond between collector conduits and the collector plate.

if you have a limited area for collectors. Otherwise cost per Btu (British thermal unit) collected is a more important factor.

The Storage System. Once the heat has been picked up by the heat-transfer medium it is carried off to storage. If the transfer medium is water, it's a simple matter to use water as the storage medium, too. Place a large tank of water in your basement. Run the water through the collectors via pipes, then back to storage. With each pass through the collectors the temperature of the water will rise. Soon you have a tank of hot water in your basement. When you need heat for your home you can extract it from the hot water by running the water through a heat exchanger (a series of finned pipes) and blowing air over the fins.

If the heat transfer medium is air, it's easiest to store the heat in a big bin of rocks. Place the bin in your basement, and run the air through the collectors and bin via ductwork. This will heat the rocks. To extract the heat for use, just run the air from your home through the bin.

The Distribution System. After extracting heat from storage, you can distribute it throughout your home quite simply. If you are using an air system, just blow the warmed air through

BASEMENT ROCK BIN

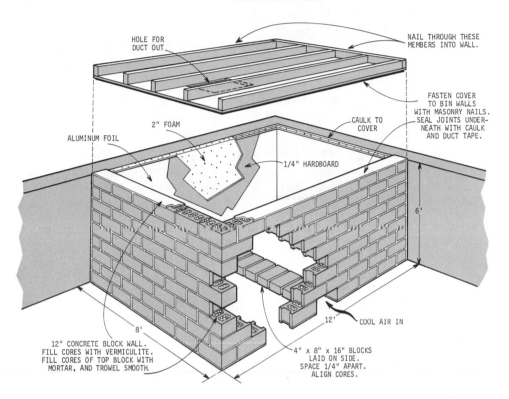

HOLE FOR
DUCT OUT

NAIL THROUGH THESE
MEMBERS INTO WALL.

FASTEN COVER
TO BIN WALLS
WITH MASONRY NAILS.
SEAL JOINTS UNDER-
NEATH WITH CAULK
AND DUCT TAPE.

2" FOAM

CAULK TO
COVER

ALUMINUM FOIL

1/4" HARDBOARD

6'

8'

12" CONCRETE BLOCK WALL.
FILL CORES WITH VERMICULITE.
FILL CORES OF TOP BLOCK WITH
MORTAR, AND TROWEL SMOOTH.

12'

COOL AIR IN

4" x 8" x 16" BLOCKS
LAID ON SIDE.
SPACE 1/4" APART.
ALIGN CORES.

Heat storage for an air system is best accomplished in a basement bin filled with rocks. The rocks can range between fist size down to about 1½ inches in diameter. If they are too large they will provide less area for heat exchange. This will force your collectors to operate at a higher temperature and, as a result, lower efficiency. If the rocks are too small, they will pack tightly together and restrict the passage of air through the bin.

Whatever the size, the rocks should be washed before you dump them into place. Dirty rocks may circulate dust through your home. Harry Thomason has found that the rocks actually act as a dust trap. Dust settles out on top of the uppermost layer. But if the rocks are exceptionally dusty when you place them, it may take some time for the air to clean them.

The bin as shown is six-feet high. This gives you room between the bin and the basement ceiling so that you can climb up on the bin while nailing the cover down. If your ceiling is lower you may want to use one less course of blocks for bin walls to give more working room. Fastening the cover down is easi-est with a gunpowder-powered fastening tool such as a Ramset.

The system shown here for distributing incoming air through the bin is patterned after a design by Harry Thomason. The principle is patented; to use it you'll need a license from Edmund Scientific., 150 Edscorp Bldg., Barrington, NJ 08007.

Note: By placing a preheating tank for your domestic water supply inside this bin, you can heat water before it goes to your water heater. This will cut winter heating costs for water, but it won't help in summer since the bin won't be heated.

The bin as shown utilizes two basement walls to cut costs. You can, however, build the bin freestanding. To simplify loading the bin with rocks, build it under a basement window. Rocks can be dumped in through a chute, cobbled together out of wood.

Insulation alternative. Since the existing basement walls used for the bin are inaccessible on the outside, they are insulated on their inner faces. The other two walls may be insulated on their outer faces if you like. This will add those two walls to the storage mass.

conventional warm-air heating ducts. If you have a liquid heat-transfer medium you might be tempted to run it through ordinary hot-water baseboard units. But this is a poor idea. These baseboard units are designed to work with very hot water—at least 140 degrees F. Your water from storage will rarely be that hot, so baseboard units will do a poor job of heating. That's why it's better to run the hot water through a heat exchanger and blow air over the fins. This allows you to get more useful heat out of your storage water. And it allows you to use the water down to a lower temperature.

That's important because your storage system is designed to hold a supply of heat for use during the night or on cloudy days when your collectors aren't collecting. As you extract heat from storage, the temperature of the water in the tank will drop. If you used baseboard units, you'd get very little heat from the water, even though it might be 125 degrees. There would still be a lot of heat in the water, but it would be useless because your baseboard units couldn't extract it. By going to a heat exchanger and blower unit, you could still extract heat when the water temperature had dropped below 100. Thus a heat exchanger might still be warming your home a few days after a baseboard system had become useless during a long cloudy spell.

One other point. We've already mentioned that collectors are most efficient when run at low temperatures. If you insist on using baseboard units with hot water, you'll be forcing your collectors to run at higher temperatures—and lower efficiencies.

The Backup System.   There will come a time when your solar heating system won't be able to provide you with all the heat you need to keep your home comfortable. You'll hit a long spell of cold, cloudy weather. You'll use up all the heat in your storage system. That's where your backup heating system comes in. In its most economical form, a backup system can be a good woodburning stove. In most installations, however, the backup turns out to be a complete conventional heating system—either a gas or oil furnace, or electric baseboard heat.

Why can't a solar heating system handle your entire heating load, without the need for a backup? It can, but it will cost too much. A solar heating system designed to meet 100 percent of your heating needs must be sized to meet the worst combination of weather conditions imaginable. And even though these conditions may occur only a few times a year, you may end up quadrupling the cost of your solar installation to meet them. That's why most solar heating systems today are sized to give you about 75 percent of your total heating needs.

AIR-COOLED VS. LIQUID-COOLED COLLECTORS

We've already mentioned that there are two basic types of collectors: air-

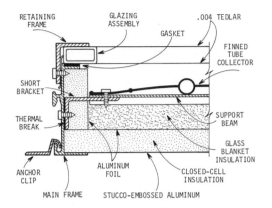

REYNOLDS THOREX COLLECTOR (Cross Section)

The Reynolds Thorex solar collector measures 4 × 8 feet in area, and is 3⅝ inches thick. The collector plate is a single piece of integrally-finned extruded aluminum. For corrosion protection, Reynolds recommends the use of an inhibitor in the collection water. Virco-Pet 30 by Mobil is their first choice, used in a concentration of .5 to 1.0 percent by volume. In addition, the collector loop should be drained, flushed, and refilled with fresh solution every two years at minimum. The collector panel is glazed with two layers of Tedlar plastic film. In tests conducted by DuPont, this plastic has retained at least 50 percent of its tensile strength after 10 years exposure in Florida.

ers made on a small enough scale to fill the bill in cooling the average-size home.

Water-cooled collectors have other advantages. The pipes that carry the water from the collectors to storage are smaller than the ducts required to move air. This means you can route them through smaller spaces with less effort than you'll need to run ductwork. The power required to pump the water through the system is also less than the power required to blow large volumes of air through an air system. And the heat transfer coefficient between a liquid and a heat exchanger surface is higher than that between air and a heat exchanger. This means you can use a smaller—and cheaper—heat exchanger when you work with a liquid system.

and liquid-cooled. Which type is best? It all depends. Each type has its own set of advantages and disadvantages.

Liquid-cooled Collectors. This is the more versatile of the two collector types. With a water-cooled collector you can heat your home, your swimming pool, your domestic hot water supply. You can also use the collectors to power an absorption type air-conditioner to cool your home in summer, at least in theory. Right now, however, there are no absorption cool-

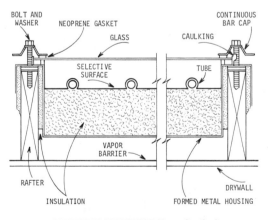

SUNWORKS COLLECTOR (Cross Section)

The Sunworks water-cooled collectors come in two types. The flush-mounted design shown here goes right between roof rafters and doubles as the roofing material. The surface mounted collector is similar to the Reynolds Thorex collector and can be installed over existing roofing materials.

REVERE COPPER COLLECTOR

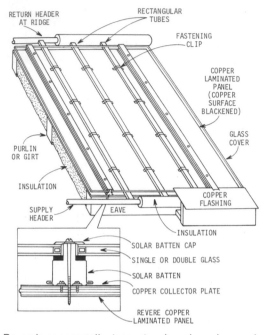

RETURN HEADER AT RIDGE

RECTANGULAR TUBES

FASTENING CLIP

COPPER LAMINATED PANEL (COPPER SURFACE BLACKENED)

GLASS COVER

PURLIN OR GIRT

INSULATION

SUPPLY HEADER

EAVE

COPPER FLASHING

INSULATION

SOLAR BATTEN CAP

SINGLE OR DOUBLE GLASS

SOLAR BATTEN

COPPER COLLECTOR PLATE

REVERE COPPER LAMINATED PANEL

Revere's copper collector system is an ingenious and durable approach to solar design. The collector plate is a sheet of copper fastened to plywood. This plywood/copper sandwich can serve as the roofing material in new construction. Under the plywood — inside the house — goes insulation which can double as both house and collector insulation. To the top of the copper sheet are fastened rectangular copper water passages, secured with clips and thermal cement. A system of battens supports the glazing necessary to prevent thermal losses from the front of the collector. Copper flashing encloses the collector to make it weathertight. The water passages are fed water through manifolds at the bottom of the collector; a second run of manifolds runs across the top of the collectors to gather water for return to storage. Revere has developed the special fittings required to connect the rectangular water passages to standard cylindrical pipes.

FALBEL OPTICAL SYSTEM

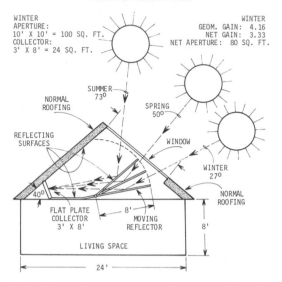

WINTER APERTURE: 10' X 10' = 100 SQ. FT. COLLECTOR: 3' X 8' = 24 SQ. FT.

WINTER GEOM. GAIN: 4.16 NET GAIN: 3.33 NET APERTURE: 80 SQ. FT.

SUMMER 73°

SPRING 50°

NORMAL ROOFING

WINTER 27°

REFLECTING SURFACES

WINDOW

NORMAL ROOFING

40°

FLAT PLATE COLLECTOR 3' X 8'

MOVING REFLECTOR

8'

LIVING SPACE

24'

Falbel Energy Systems uses a unique pyramidal optical system to concentrate sunlight gathered over a large area onto a smaller solar collector. Since the cost of building reflectors is lower than that of building a collector, the overall system costs less. As shown here, the collector is located inside the attic. A large window covers most of the south face of the roof. The inside of the attic is lined with aluminized Mylar film, a highly reflective material. A movable reflector can be set at different angles throughout the year to maximize the amount of sunlight reflected onto the collector. This system can concentrate sunlight falling on the collector by a factor as high as six. The collector can be air or water cooled.

Now for the bad news. Liquid systems can leak, and when they do they can cause real trouble. What would you rather put up with, a small air leak from the ductwork in your attic, or a stream of water seeping through your ceiling and running down your walls? Liquid cooled collectors are usually more expensive. They can suffer from corrosion. In cold climates the water in them can freeze at night or in cloudy weather. You can solve this problem by draining the collectors at night. Or you can run a mixture of antifreeze and water through the collectors. Using antifreeze in your complete water system—including the storage water—would be very expensive. It might require 750 to 1,000 gallons of antifreeze. So in practice, you'd run a closed loop of antifreeze through your collectors and down to the storage tank. At the tank you'd have a heat exchanger—usually a coil of tubing inside the tank—to pass the heat from the antifreeze solution to the storage water. Of course all this adds to the expense.

Air-cooled Collectors. If you are interested solely in space heating, air-cooled collectors have advantages. You can take hot air directly from the collectors or from storage and duct it wherever you need it. Air-cooled collectors are simple to adapt to an existing hot-air heating system, which then becomes your backup system. Air leakage is not the grave problem that water leakage is. Corrosion is not much of a problem. Your storage system—a concrete bin of rocks—will be cheaper than a steel tank of water in most cases. Your collectors will be cheaper than their water-cooled counterparts.

And air systems can be more efficient. Remember that it becomes difficult to extract heat from water once it drops much below 100 degrees. At that point you'll still have a lot of heat in storage, but you can't get it out. The effort is useless. But with

an air system you can still get useful heat out of a bin of rocks, even with its temperature way down at 75 degrees. So with air you can run your whole collector system at lower temperatures (and higher efficiency), and you can get more useful heat out of your storage medium.

One last advantage to an air system: If you decide to build your own collectors, you'll find the air-cooled type is easier to make.

After adding up the pros and cons of each type of system, I favor air-cooled collectors for space heating. As noted, to get the most out of any solar heating system, you'll probably have to use a forced-hot-air distribution system. As long as that's the case, you might as well use air-cooled collectors. They'll mesh nicely with your distribution system, and they should cost less. Over the years they'll probably give you less trouble than water-cooled collectors with the possible exception of all-copper, water-cooled collectors. (There is still some question about the corrosion resistance of aluminum collectors over the long haul.)

If you decide to use solar energy to heat your water, go with a separate water heating system, equipped with its own set of water-cooled collectors. But for space heating, favor air-cooled collectors.

Oddly, most makers of solar hardware don't agree with this philosophy. The overwhelming majority of them manufacture water-cooled collectors. It's easy to understand why a

DO-IT-YOURSELF LIQUID COLLECTOR

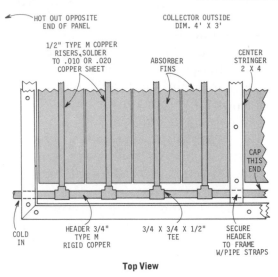

Top View

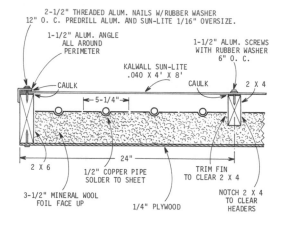

Cross Section

The header/riser collector panel is ideally suited to thermosiphon applications. You can build it yourself to collect heat for a water heater. It's designed for easy construction with a minimum of waste, yet it has high efficiency due to copper construction. As shown here, it is single glazed with Kalwall Sun-Lite. Double glazing is not worth the time and expense in most parts of the country, especially for summer use only.

To build the collector, start with the frame. Glue and nail up the perimeter framing using good 2x6 stock. Use redwood if you like, but pressure-treated fir will last as long. Ordinary fir treated with pentachlorophenol wood preservative will last almost as long, but don't apply it until all parts are assembled or it might interfere with the glueing process.

After the perimeter frame is together, nail and glue on the plywood bottom. Then cut the center stringer to length, but don't install it. It goes in place only after the collector has been installed in its frame.

Next, cut all your copper pipe to length. The risers will run about 92 inches long, but test fit them with Tees in place, leaving about ¼-inch clearance at each end of the Tees to allow the pipes room to expand when they heat up. The lengths of header pipe will run about 4¼ inches long, but again, check them for fit. When all parts are cut, assemble them dry and check the whole array for fit.

Form the copper fins to accept their risers as shown in the bottom drawing. These fins should be about 1½-inches shorter than their risers or about 90½ inches long. When the fins are formed, clean them carefully with steel wool so your next step—soldering them to the risers—will go easily. Clean the risers with steel wool, too, wearing cotton gloves to keep oil

off the copper. Then coat pipes and fins lightly with soldering paste and solder the tubes into the grooves formed in the fins. A propane torch and ordinary solid 50/50 solder will do the job. One of the new MAPP torches will make the job go faster. Try to lay a good fillet of solder all along the joint. This will assure a good thermal bond, and will boost efficiency.

When the fins are all attached, clean the ends of all pipes and the insides of all Tees. Coat with paste, assemble the panel dry, and sweat solder all joints. Be sure the fins are aligned properly. Scrub the completed panel with hot water, cleanser and a brush. Then hook it up temporarily to a pressurized water line and test for leaks. Resolder any leaking joints.

Paint the collector with a good high-temperature paint. Sapolin makes a flat black barbecue paint that holds up well. Or you can order a can of black absorber coating from Kalwall for about $5 a quart. While the paint is drying, slip the insulation into the collector frame. Staple the foil facing to the faces of the frame so the insulation can't slide down when the collector is installed at the proper slant. Use duct tape to seal the joints between pieces of insulation.

Paint dry? Slip the panel into the frame, sliding the inlet and outlet nipples out through their holes. Secure the panel in place with pipe straps run around the headers and nailed to the collector frame.

Last step is installing the glazing. Nail the center stringer in place, notching it if necessary to clear the headers. Lay a fat double bead of Tremco Mono caulk all around the top edges of the frame, and down the top of the stringer. Place the glazing material and fasten in place with washer-headed nails around the perimeter, and with screws down the stringer. Note

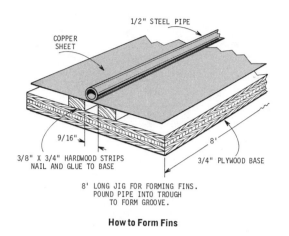

1/2" STEEL PIPE

COPPER
SHEET

9/16"

3/8" X 3/4" HARDWOOD STRIPS
NAIL AND GLUE TO BASE

8'

3/4" PLYWOOD BASE

8' LONG JIG FOR FORMING FINS.
POUND PIPE INTO TROUGH
TO FORM GROOVE.

How to Form Fins

that the perimeter of the frame is capped with alumi-num angle. This reinforces the Sun-Lite along the edges, and helps seal out moisture. Run a bead of caulk under the aluminum before you fasten it down.

Despite your best efforts, some moisture may ac-cumulate inside the collector frame. So drill a few ¼-inch weep holes through the plywood bottom. Paint the outside of the collector if you want. That's it!

large copper-producing corporation like Revere would turn to water-cooled hardware; copper is a natural in water carrying applications. It's less easy to understand why alumi-num manufacturers don't build air-cooled collectors. Aluminum is ide-ally suited to this application, and much less suited to water carrying applications. Corrosion is such a problem that most makers recom-mend running a solution of distilled water (or rain water) and automotive antifreeze formulated for aluminum-block engines such as the Chevrolet Vega through water-cooled alumi-num collectors. Even so, it's a good idea to add a getter column to any aluminum system running on liquid

coolant. (A getter column is a section of pipe in the collector loop contain-ing a length of rolled-up aluminum screening. It is placed upstream of the aluminum collectors so that any corrosive elements in the circulating liquid attack the screening and be-come neutralized before they reach the collectors.)

Despite the scarcity of air-cooled collectors, the systems that are avail-able—Solaron or Sunworks, for ex-ample—are of high quality. So don't let the scarcity of manufacturers sway your decision. If you decide to use air-cooled collectors, good ones are available. (Sources for equipment are listed at the back of this chapter.) And don't forget you can always make your own.

MATERIALS FOR SOLAR COLLECTORS

Whether you buy your collectors or build your own, it pays to know about the materials that go into them. Which materials perform the best, last the longest, and provide the most value for your money? Which are eas-iest to work with?

Absorber Plates. The most common materials here are copper, aluminum, plastic, steel, and occasionally a few other materials such as glass or cloth. Which material you choose depends in part on collector type. By far the best and most reliable material for water-cooled collectors is copper. It provides the best heat transfer. It is the most corrosion resistant. And if you build yourself, copper is easy to work with. Its only real drawback is its cost—roughly twice that of aluminum.

Aluminum may be cheaper, but it has drawbacks. Biggest problem—as mentioned earlier—is corrosion, at least in water-cooled collectors. But even the antifreeze designed for aluminum block car engines and aluminum corrosion inhibitors eventually breaks down into acidic compounds that can worsen corrosion. And if you add other metals to your system (copper pipes, bronze pump parts, etc.) the chances for galvanic action increase even more.

One maker of solar collector plates—Energex—dropped its line of aluminum collectors and went over entirely to copper plates. They just couldn't solve the problems of corrosion over the long haul. To provide economical service, collectors must be able to last at least 20 years. Still, other big manufacturers continue to use aluminum. Best advice is to stay away from aluminum in water-cooled collectors whenever possible, in favor of copper.

In air-cooled collectors, however, aluminum is the answer. Corrosion is no longer a problem. Aluminum's lower price and weight make it a good choice. And in air-cooled collectors, aluminum is easy to work with, an important factor if you plan to build your own.

Note: All the above cautions on the use of aluminum in water-cooled collectors apply to conventional collec-

DO-IT-YOURSELF AIR-TYPE COLLECTOR

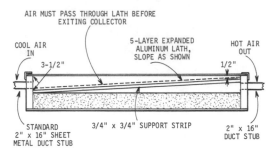

Side View

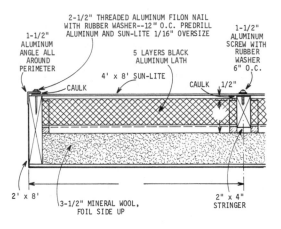

End View

This easy-to-make air-type collector features an absorber matrix of expanded aluminum lath. You can buy the lath from Kalwall, also the source for the Sun-Lite cover plate. The sheet metal duct stubs come from any local sheet-metal shop. All other parts are available at most hardware and lumber outlets.

Start by building the frame. Pressure-treated lumber is a good idea for the main frame components. Nail and glue the frame together, then add the back. Install the duct stubs at the ends of the frame before you put in the center stringer. The stringer will bisect the duct stub openings, obstructing them to some extent, but this will not seriously affect air flow.

Note that the matrix of five aluminum lath panels slopes from one end of the collector to the other. This forces the air to pass through the lath on its way from one end of the collector to the other. It also puts all the air that passes through the matrix (and has thus been heated) behind the matrix and isolated from the cover sheet. This helps prevent the hot air from losing heat to the cover sheet.

The five layers of lath rest on wooden supports cut from ordinary one-inch pine. Nail these to the sides of

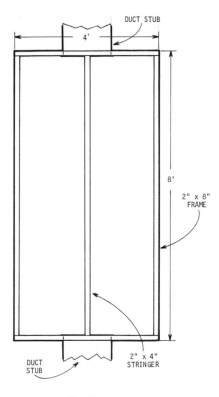

Top View

the side members of the frame, and to both sides of the center stringer. Note that the lower ends of the supports fastened to the stringer will hang down one-half inch below the bottom edge of the stringer. The lath as it comes from Kalwall will be 27 inches wide, so you'll have to trim it a bit to fit the collector box. Heavy shears or tinsnips will do the job."

The cover sheet goes on last, after the insulation has been placed, and the lath has been tack nailed or stapled to its support strips. Use Filon nails (with washers under the head) around the perimeter of the frame, and use screws down the center of the stringer. Predrill for both nails and screws, making the holes $1/16$-inch oversize to allow for expansion and contraction of the aluminum angle and the Sun-Lite cover plate. Use a good elastic caulk such as Tremco Mono between the plate and the frame and stringer, and also underneath the aluminum angle. Insulate all ducts connecting the panels and leading to storage.

Insulate the bottom of the box with $3\frac{1}{2}$-inch foil-faced fiberglass blankets. Two pieces 24 inches wide will do the job nicely. Staple the flanges of the vapor barrier to the frame and stringer to keep the insulation from shifting.

Over the insulation goes the heat-absorbing matrix of aluminum lath. Bind the layers together at a few points with some twists of wire. While you're doing this, try to arrange the five separate layers so that you can't see through them. This will assure maximum interception and absorption of the sunlight entering the box. Any light that does manage to pass all the way through the matrix will be reflected back toward the matrix by the foil vapor barrier on the insulation.

Air flow through the collectors should be adjusted to equal about three cubic feet of air per minute, per square foot of collector area. A single collector of this size would require a rate of flow around 120 cubic feet per minute. That sounds like a lot of air, but it's really a fairly gentle flow. It amounts to a change of the air inside the collector every six seconds.

tors with closed water passages. There is one type of water-cooled aluminum collector that does make a lot of sense. It's the design worked out by solar pioneer Harry Thomason. His collector plate is corrugated aluminum roofing, and the water simply trickles down the valleys of the corrugations. Thomason uses rainwater in his system—with no expensive additives of any kind—and figures the panels will last from 15 to 20 years.

Thicker aluminum would last even longer. But in any case, the collectors are of a fairly simple design and could be rebuilt with much less effort than other designs once the aluminum begins to deteriorate.

Plastic? Its main place is in low-temperature solar applications such as swimming pool heaters. There the temperature inside the collector may be only a few degrees above that of the outside air. There is not enough heat to accelerate the breakdown of the plastic, so service life is reasonable. Plastic also serves well in some domestic water heaters, notably Japanese designs such as the Hitachi. So far, however, it has not made its way into space heating.

Steel? This is not a bad choice for air-cooled collectors, but it's subject to corrosion in water-cooled applications.

Cover Plates. Most solar collectors need a clear cover plate to produce optimum performance. This plate slows the escape of trapped heat. The bigger the difference between the temperatures inside and outside of the collector, the more important the cover plate becomes. In swimming pool heaters, the temperature difference is very small, so no cover plate is required. The amount of heat the plate might save is more than offset by the expense of the plate, and by the fact that it will reduce the amount of sunlight reaching the collector by about 10 percent.

When the temperature difference gets up to about 30 or 40 degrees, a cover plate will improve performance. Despite the fact that this cover will intercept about 10 percent of the light before it can reach the absorber plate, it's still an asset. Why? Because now the temperature difference between the plate and the outside air is so great that the plate will lose much of its heat to the air. The cover plate slows that heat loss, and winds up saving more than it costs.

What about two cover plates? Double cover plates begin to pay off when the temperature difference reaches about 100 degrees.

What materials make the best cover plates? Glass has some advantages. So do a few plastics. Some plastics, however, just can't do the job. Let's take a look.

Glass. As long as breakage isn't a potential problem, glass is a good choice. The stuff will last forever without deteriorating or losing its transmittance. But where hail or vandals are a problem, glass makes a poor

This table prepared by Kalwall compares the properties of their Sun-Lite glazing material to ordinary glass. Sun-Lite is one of the best plastic glazing materials available for collector construction.

	Ideal	3/16″ Glass	Sun-Lite
Transmittance, solar, ASTM E424	1.00	.83 to .87	.85 to .90
Transmittance, heat (5 $_u$M to 50 $_u$M)	0.00	.1	.1
Thermal Conductivity, (BTU-in/hr-ft²-°F)	0.00	6 to 7	.87
Thermal Expansion Coef., PPM*/°F	12**	4 to 5	14
Impact Strength, SPI ball drop, ft-lbs	high	10	60
Ultimate Tensile Strength, ASTM D638, psi	high	10,000	16,000
Weight per sq ft (.040″ Sun-Lite), lbs	low	2.4	0.3
Resistance to chemicals, UV & time	excellent	excellent	very good
Handling & Cutting	safe, easy	difficult	safe, easy
Cost (in quantity from distributor), $/ft²	low	.52	.28 to .46
* Parts per million		** To match aluminum used in most collectors	

Kalwall's Solar Components Division offers a range of solar products including pumps, blowers, collectors, glazing material, solar water heaters, pool heaters. At left, is one of their preassembled, double-glazed collector covers which can also be used as a south wall building panel in passive solar construction, as shown at right. The panels are a frosty "ice clear." You can't see through them but they transmit more solar energy than glass.

choice. Glass also poses a safety hazard if your collectors are on the ground or mounted on a fence. Tempered safety glass can alleviate these problems, but at a price; it costs more, and it's impossible to cut at home, a definite disadvantage if you want to build your own collectors.

Plastics. Probably the best choice would be a quality fiberglass reinforced plastic. Note, however, that not all these products are the same.

The common garden variety, designed for patio roofs, will not hold up for long on a solar collector. They'll break down, begin to yellow, and lose their transmittance. For solar panels, be sure to use one of the special formulations containing additives that prolong life. Kalwall's Sun-Lite is such a material. While these plastics are not glass clear, they do transmit sunlight as well as glass, or even a few percentage points better.

Oddly, the somewhat frosty appearance of these plastics doesn't impede the passage of solar radiation. And in fact the frosty quality may help distribute the light evenly over the face of the collector plate.

These plastics will generally cost you less than 3/16-inch glass, depending on the grade you buy. For example, Kalwall's Sun-Lite runs from about 28 to 46 cents a square foot, while glass·costs over 50 cents. When it comes to working and handling, plastic has a clear advantage: It's lighter by a factor of eight. It cuts and drills with no problems. It's much more impact resistant, and when it does break it presents no safety hazard. Drawbacks? Expected life of the best grades is about 20 years. That's for a premium grade. A standard grade might last you about seven years. In situations where the cover is not subject to physical damage, glass would last forever, and would probably make a better choice.

Other   plastics.   Polyethylene, vinyl, Mylar, and Tedlar have all been tried, and most have failed. The problems are basically two: First, thin films of plastic are subject to wind damage. They eventually tear or shatter. Second, they can't hold up under the ravages of ultraviolet radiation. They get brittle, lose their clarity, and break down. Polyethylene might last a few months, the others somewhat longer. But they all go within a short time, except for Tedlar, which can last 10 years or more.

In collectors with two cover sheets, plastic films have longer lifespans. For example, Garden Way puts a four-mil film of Tedlar into their collector, placing it a half-inch below the outside cover plate of glass-reinforced plastic. The outer sheet protects the Tedlar from wind and radiation damage and increases its life significantly.

Insulation for Solar Collectors.   For good efficiency, a solar collector must be insulated. A barrier of insulation behind the collector plate slows heat loss in that direction. Most collectors are not insulated around their edges. The insulation would take up space that could otherwise be absorbing solar energy, so the space is more effectively used as collector plate.

Insulation for a solar collector has to be able to withstand some fairly high temperatures. A good flat-plate collector with double glazing might reach temperatures over 300 degrees F. under some conditions. That's enough heat to melt most plastic foams. Mineral wools, therefore, make a better choice. Their weakness is that they provide less insulation value per inch than the foams do. So they result in a bulkier collector. One way around the problem is to use mineral wool insulation directly behind the plate, then back it up with an inch or so of foam. The wool provides enough insulation to protect the foam, and the foam adds an extra measure of insulation in a compact package.

Collector Frames. Most commercially made collectors have metal frames. Industry is well equipped to fabricate frames from metal, and metal will hold up well over the years. So industry's choice is a good one. If you want to build collectors, though, metal may not be your best choice. It's difficult to handle in the average home shop. If you can bargain with a local sheet-metal fabricator, you may be able to have your frames built for you. If not, the logical choice is wood. But will wood hold up? Take a look around and you can see houses made of wood that have lasted hundreds of years. But that wood isn't subject to the severe temperature fluctuations you'll find in a solar collector. There has been relatively little testing of wood in solar collectors. Harry Thomason has found that ordinary 2x4s used in some of his early collectors would last only about five years. But redwood or pressure-treated lumber (Kopper's Wolmanized lumber, for example) should be able to last much longer than that. One way to prolong the life of a wooden collector frame would be to insulate the frame. As mentioned earlier, insulating the frame will cut into space that could be better used to collect solar energy. But in this case, you might be willing to trade a little collector area in order to lengthen the life of your wooden collector parts.

SOLAR WATER HEATING

Probably the best first step for anyone interested in cutting costs and energy consumption with solar heat is to install a *water* heating system. A solar heater for domestic hot water is an easy first step. It's relatively cheap — $1,000 to $2,000 — for a commercial unit, even less if you make your own. And time to payback runs an average of about six years. A good homemade unit might pay off in four years; an expensive commercial unit may take as long as ten, counting installation costs.

One last reason for starting with solar water heaters: Unlike solar space heaters, they are easy to tie into existing homes. No matter how poorly insulated, weatherstripped, caulked, or glazed your home might be, a solar water heater will still do its job. You can't say the same for solar space heaters. And it is important at this point in history to begin the changeover from fossil fuels to solar power as soon as possible.

Types of solar water heaters. The simplest water heater can circulate its own water and regulate its own flow, all without the need for pumps or thermostats. It requires no outside pressure because it delivers heated water by gravity feed. At the other end of the spectrum, you'll see water heaters with pumps, differential thermostats, heat exchangers, and so on. While the first system might cost as little as $300 complete, the more complicated might cost you close to $2,000. Still, there are good reasons why you might choose one type over the other. So let's take a look.

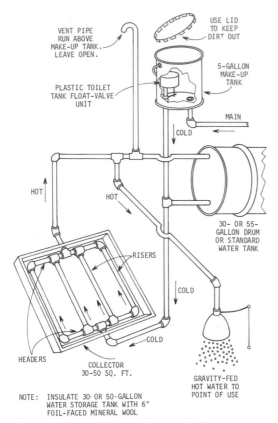

VENT PIPE
RUN ABOVE
MAKE-UP TANK.
LEAVE OPEN.

USE LID
TO KEEP
DIRT OUT

5-GALLON
MAKE-UP
TANK

PLASTIC TOILET
TANK FLOAT-VALVE
UNIT

MAIN

COLD

HOT

HOT

30- OR 55-
GALLON DRUM
OR STANDARD
WATER TANK

RISERS

COLD

HEADERS

COLLECTOR
30-50 SQ. FT.

COLD

GRAVITY-FED
HOT WATER TO
POINT OF USE

NOTE: INSULATE 30- OR 50-GALLON
WATER STORAGE TANK WITH 6"
FOIL-FACED MINERAL WOOL

Simple Thermosiphon. This type of water heater is a marvel of simplicity. It may not be suited to all homes or all climates, but where it does work it will serve you well. Take a look at the accompanying drawing to see how a thermosiphon water heater works.

A solar collector panel should be mounted at the proper angle for your latitude. Basic collector construction consists of header pipes at the upper and lower edges of the collector, with risers running between the two. This collector is connected via two pipes to a hot water storage tank, located a minimum of two feet above the collector. One pipe from the top of the collector connects to the top of the tank. The other pipe, running from the bottom of the collector, connects to the bottom of the tank.

A simple thermosiphon water heater for a summer home should be laid out as shown here. This system operates entirely on its own power. The thermosiphon principle circulates the water through the collector, and gravity feeds the heated water to its point of use. Since the unit is designed only for summer use, you'll be able to get by with a minimum of collector area. Just 30 to 50 square feet of single-glazed metal collector will do the job.

You can buy the collector, or make it yourself. In either case, be sure the collector is designed to work well in a thermosiphon system. For best results, the collectors should have header and riser construction, with the headers at the top and bottom of the collector, and the risers running vertically between the headers. Serpentine-type collectors will not generate enough pressure to circulate the water. Neither will coil-type collectors. If you buy, stick to copper collectors, for aluminum may not hold up well circulating untreated water. The Sunworks collector is a good choice.

If you build your own collector, the design for the header/riser panel shown earlier in this chapter will do the job nicely, and it's easy to put together. Single glazing is your best bet for a summertime collector. It will probably give better performance than a double-glazed collector, and it will naturally cost less to make. For economy and ease of construction, con-

sider Kalwall's Sun-Lite, a glass-reinforced, plastic sheet material. A piece 10 feet long and four feet wide will cost about $20, and will provide enough material for a complete installation. Kalwall also sells Tremco's Mono, an excellent caulk for sealing this glazing material to the collector frame.

Plumbing should be CPVC plastic pipe. This stuff is easy to work with. It cuts with a saw, goes together with cement, costs less than other types of pipe, and will retain heat better, too. If you keep the runs between the collector and the storage tank short (10 feet or less) you can get by with ¾-inch pipe. If the runs are longer, you'll be wise to switch to one-inch because smaller diameters may constrict the flow of water through the thermosiphon loop. Trouble is, you may not be able to find plastic pipe in a diameter over ¾-inch, so you may have to switch to copper. Copper will cost more and it's harder to handle. The key is to keep the thermosiphon runs as short as possible.

Note: Construction is a problem only in the thermosiphon loop, so you can get by with ½-inch plastic pipe in all your other runs.

With a thermosiphon heating system, insulation of your hot water lines is very important. Use Arma-Flex insulation—sold at plumbing outlets—or wrap the pipes with ordinary 3½-inch foil-faced mineral wool, foil side out.

As the sun heats the water in the collector, the hot water rises. It runs up the risers to the top header and on up to the top of the storage tank. Meanwhile, cold water from the bottom of the storage tank runs down through its pipe to the bottom of the collector panel. This water is then heated and begins to rise. Result? Water continually circulates through the panel and up to the tank. The hottest water is always at the top of the tank, the coolest settles to the bottom where it can be picked up and piped off to the collector for heating.

As long as there is sun shining on the collector, the thermosiphon effect continues. As soon as the collector can no longer raise the temperature of the water in storage, the water stops circulating. In effect, you have an automatic differential thermostat controlling the flow. Warning: If the entire storage tank is not located above the collector, the thermosiphon principle can operate in reverse when the sun goes down. That is, as the water in the collector cools, it will fall to the tank and set up reverse circulation. Water passing through the collector will cool off and soon all the water in storage will lose its heat. That's why the tank has to be at least two feet above the top of the collector.

Okay, to make use of the hot water you can run a pipe from the storage tank (near the top where the water is hottest) to the point of use. If the collector and tank are mounted on the roof, gravity flow will deliver the water anywhere in your home. To keep the whole system topped off with water, you hook in a small makeup tank. A toilet tank valve allows water at house pressure to fill this tank, then shuts it off. Now as water is used, water from the makeup tank enters the system via gravity. Finally, to keep the whole system at atmospheric pressure, and to allow ex-

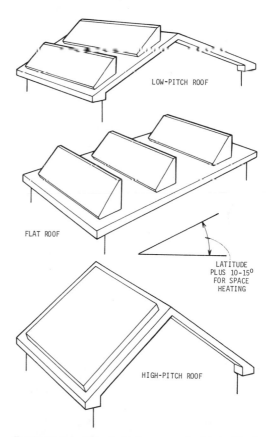

LOW-PITCH ROOF

FLAT ROOF

LATITUDE PLUS 10-15° FOR SPACE HEATING

HIGH-PITCH ROOF

Roof angles may not always coincide with the optimum mounting angles for solar collectors. The drawings here show a few ways to mount collectors at the correct angle. A good rule of thumb for year-round space heating is to set the angle at latitude plus 10 or 15 degrees.

cess pressure to dissipate, a vent pipe runs upward to a point higher than any other point in the hot water system. There the pipe terminates, its end left open and uncapped.

That's all there is to it. This rig will do a great job as long as outside air stays above freezing. But in freezing weather the pipes can freeze up and burst. A heat exchanger can prevent such a disaster.

Thermosiphon with Heat Exchanger.   In this system, everything is the same as in the system above, with one exception: a heat exchanger. And adding a heat exchanger is no big deal. Here's all you have to do:

Where the hot-water pipe from the top of the collector enters the top of the tank, connect it to a 40-foot spiral of copper tubing that runs towards the bottom of the tank. There connect the tubing to the cold-water outlet running back to the collector. The copper spiral provides a large surface for transmitting its heat through its walls to the fresh water in the tank.

Now you can fill the collector and its run of pipe and tubing with a mixture of water and antifreeze. This will prevent any freeze-up problems, and will in no way affect the thermosiphon principle. The water will circulate as usual, rising through the collector as it heats up, falling downward through the heat exchanger coil as it gives up its heat to the water in the tank. As long as you keep your tank insulated and inside the house, you shouldn't have freezing problems.

Warning: Ordinary automotive antifreeze—ethylene glycol—is toxic. If a leak develops in the heat exchanger coil, the antifreeze can enter your hot water supply. There are two ways around this problem: Use propylene glycol antifreeze (it's nontoxic) or modify your heat exchanger. Instead of running the coil inside the tank, spiral it around the outside of the tank. Assuming the tank is metal, heat transfer will be only slightly reduced, especially if you bond the tubing to the tank with a thermal cement. Then you can insulate the tube/tank to prevent heat loss, or you can sacrifice some water temperature by letting the tubing provide a modest amount of space heat.

Okay, this system prevents the freezing, but it still delivers your hot water via gravity feed. You may not like this arrangement. It is not compatible with most residential plumbing systems. There is no way to hook an unpressurized source of hot water into a normal pressurized line. So you can't combine any of the above systems with existing hot water lines. And you can't use your conventional water heater as a backup—unless you want to install a dual system of hot water lines. You could use solar-heated water at gravity pressure from one set of pipes. If that supply runs out, you could turn to your normal pressurized system for backup. This approach might be tolerable if you wanted to use your solar-heated water only at a limited number of points. But here's a better bet:

SUNWORKS SOLECTOR

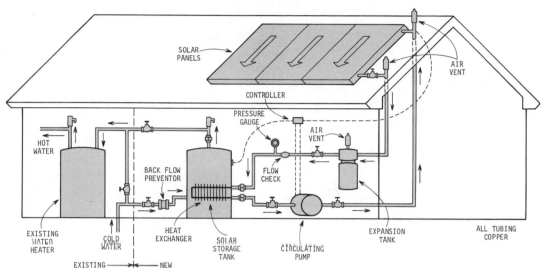

Sunworks' collector pumps water/antifreeze solution through rooftop collectors and dumps collected heat via a heat exchanger inside an insulated storage tank. Since the exchanger could spring a leak, allowing water/antifreeze to enter the potable water supply, nontoxic propylene glycol antifreeze is used. A differential thermostat controls the pump that circulates the collecting fluid. As long as the collectors are hotter than the storage tank, the pump will run and collect the heat.

This is a first-rate system. It has copper collector panels, proper venting, backflow prevention, and an expansion tank. The collectors are single-glazed because research has shown a single-glazed panel will collect more energy over the course of a year than a double-glazed panel will. Thus a year-round collector for water heating (as opposed to a winter-only system such as a space heater) performs best with single glazing. A 60-square-foot collector system, will serve a family of four. Price: $1,600 installed.

Run your collector and heat exchanger loop through a pressurized tank, a tank fed by water at normal house pressure. If you can locate the tank above your collector, the thermosiphon principle will still work, and you won't need extras such as pumps. Now that your hot water is at house pressure, you can do more with it. For example, instead of using it directly, you can now pipe it to your conventional water heater. If the water is already at working temperature, it will go right through the con-

ventional heater without requiring additional heating. But if the sun can't provide enough heat to do the whole job, your conventional heater can make up the difference. This technique of preheating your hot water supply via the sun is the most suitable for adapting to existing plumbing systems. It always delivers hot water at a constant temperature. If the sun can't provide water at 140 degrees (or whatever temperature you use your hot water) under marginal sun conditions, it can at

DO-IT-YOURSELF HOT-WATER SYSTEM

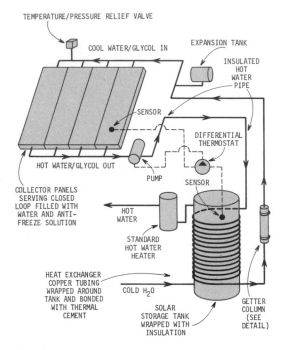

TEMPERATURE/PRESSURE RELIEF VALVE

COOL WATER/GLYCOL IN

EXPANSION TANK

INSULATED HOT WATER PIPE

SENSOR

DIFFERENTIAL THERMOSTAT

HOT WATER/GLYCOL OUT

PUMP

SENSOR

COLLECTOR PANELS SERVING CLOSED LOOP FILLED WITH WATER AND ANTI-FREEZE SOLUTION

HOT WATER

STANDARD HOT WATER HEATER

HEAT EXCHANGER COPPER TUBING WRAPPED AROUND TANK AND BONDED WITH THERMAL CEMENT

COLD H₂O

SOLAR STORAGE TANK WRAPPED WITH INSULATION

GETTER COLUMN (SEE DETAIL)

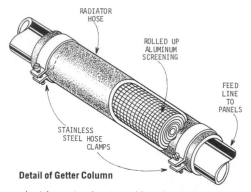

RADIATOR HOSE

ROLLED UP ALUMINUM SCREENING

FEED LINE TO PANELS

STAINLESS STEEL HOSE CLAMPS

Detail of Getter Column

This drawing shows the best way to build your own solar hot-water system so that it will interface with a typical existing hot-water system. To avoid freeze-ups, a pump circulates a solution of water and anti-freeze through the collectors and the heat exchanger coil. If you use a copper collector, you can use ordinary tap water without concern for corrosion. But if you use an aluminum collector, you'll have to use rainwater, distilled water, or deionized water in your glycol solution—plus a getter column. Placed upstream of the collector, this getter column contains a length of rolled-up aluminum screening that helps neutralize corrosive elements.

Note that the heat-exchange coil wraps around the outside of the tank, not the inside. This prevents the possibility of toxic ethylene glycol leaking into your potable water supply. For best heat transfer, bond the coil to the tank with a thermal cement. One such material is Thermon, available from A-Z Solar Products.

How big should the storage tank be? Somewhere around 150 gallons is a good size. If your tank holds much more than that, the collectors will have trouble raising the water to a workable temperature (120 degrees F. or higher). If it holds less, the water will overheat under some conditions, the efficiency of your collectors will drop, and you won't have enough storage for more than one cloudy day. You may want to buy a ready-made tank, preinsulated and equipped with a heat exchanger. A-Z sells them in 40- and 65-gallon sizes for about $200 and $300. They also sell

an electric water heater with a heat-exchange coil built into it. The heat exchanger hooks up to the solar collectors. When the sun provides enough heat, the heater's electrical element stays off. When the sun can't do the job, the electric element backs it up. Price is about $350. Neither of these ready-mades will provide cloudy-day storage, however.

Plumbing should be copper, or CPVC plastic. Do not use ordinary PVC pipe, make sure you get CPVC (chlorinated polyvinyl chloride). Ordinary PVC can't take the high temperatures inherent in a hot water system; it's for cold water plumbing only. If you do use CPVC, and you should with aluminum collectors, adapters are available to make the transitions to the copper tubing heat exchanger, and to the fittings at the pumps and collector panels. Whatever type of plumbing you use, be sure to insulate your pipes. This is most important on the hot-water run from collectors to tank, but your system will also benefit from insulation on the line feeding the panels.

How much collector area will you need? In most areas of the country, about 32 square feet will do the job during the summer. But for good performance during the winter you'll need about three times as much area. You can always add extra panels at a later date if you find your system isn't giving you the service you expect, so you can start by thinking small without the fear of wasting your time and money.

Since this system places the collectors above the storage tank, you'll need a pump to circulate the glycol solution. This needn't be a large pump, but it should be capable of handling temperatures up to 150 degrees or more. The Teel hot-water-system pump is a good choice. It's about $50 from the Kalwall Corporation. Kalwall also sells the differential thermostat you'll need to control the pump. The Delta-T 100 is a good choice; it has an automatic off feature that shuts off the pump whenever the collectors drop below 80 degrees. This will prevent the circulating pump from switching on at night if you deplete all the hot water in storage. Without this feature the pump goes on if storage drops below the temperature of the collectors, even though the collectors couldn't provide any useful heat. This might happen on a warm summer night. The collectors might be at 70 degrees, the water in storage at 60.

least do part of the job, reducing the load on your conventional heater, and saving fuel. And if the sun gives out completely, your conventional heater will work as always.

One other bonus: Since your hot water supply is now under pressure, there is no chance for antifreeze in the unpressurized heat exchanger to leak into the hot water.

If you can't locate your solar hot-water tank above the collector, you'll need a pressurized-pump system.

Pressurized-pump System. Once you install your solar hot-water tank below the collector you can no longer rely on the thermosiphon principle. You'll need a small pump to circulate water through the collector. That's not all. Remember that the thermosiphon has its own built-in regulating system. It stops circulating whenever the collectors don't do anything to raise the temperature of the water in storage. A pump system has no such self-regulating feature. You'll have to add it in the form of a twin-sensor, differential thermostat. One sensor will check the temperature of the hot water in storage. The other will check the temperature of the water in the collector. If the collector water is hotter than the storage water, the differential thermostat will turn the pump on. But if the collector water is cooler than the storage water, the thermostat will shut the pump down. This prevents the hot water from circulating through the collector at night

or during cloudy spells when the collector will actually lose heat to the outside air rather than collect it.

Drain-down Systems. There's one other way to handle water heating. A thermostatically-controlled drain valve allows you to eliminate the heat exchanger and its antifreeze solution, and run potable water through your collectors. When the temperature drops near freezing, the valve opens to drain the collectors and prevent

DEKO-LABS DIFFERENTIAL THERMOSTAT SYSTEM

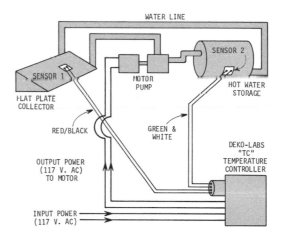

Deko-Lab's differential thermostat is one of the popular solar-control devices. Like other differential thermostats, it compares the temperatures at two different points in a solar system, then makes a switching decision based on the comparison. As this drawing shows, Sensor 1 reads the temperature of the solar collectors. Sensor 2 reads storage temperature. If the collector temperature is higher than the storage temperature by a preset number of degrees, the thermostat will turn on the pump. This circulates water through the collectors and initiates collection of heat. However, if the collectors are cooler than storage, the thermostat turns the pump off. This prevents the pump from sending warm water from storage up to the cool collectors where the water would lose heat.

HITACHI HI-HEATER

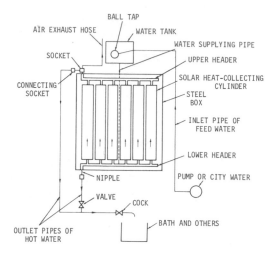

Hitachi Hi-Heater circulates potable water. Thus it is best suited for use in climates without freezing temperatures. A temperature-sensitive drain-down control could be used to evacuate the heater automatically in the event of freezing weather, but this, of course, is a waste of any energy that has been captured during the day. Garden Way and General Energy Devices sell drain-down devices. The capacity of the Hi-Heater with six absorber tanks is 55 gallons.

Note that the Hi-Heater does not operate at street pressure. It is fed via gravity from the small tank located above the heater. The heater in turn delivers water via gravity. This means the Hi-Heater will not interface with existing hot water systems. Its best application would be for a summer cottage. Price: $400.

REYNOLDS THOREX SYSTEM

The typical Reynolds Thorex installation, features drain-down to avoid freezing problems. In order for drain-down to work, the air-vent float valve must be at the highest point in the system, and the top of the sump must be level with or lower than the bottom of the collectors. In addition, the volume of air in the sump must be great enough to allow for displacement of water in all collectors and connecting piping. Even with corrosion inhibitors, use a minimum of copper materials in the system. All connections between aluminum collectors and pipes made of other alloys must be made of non-metallic fittings. The outer surface of the heat exchanger—unprotected by inhibitors—can be protected by using sacrificial anodes.

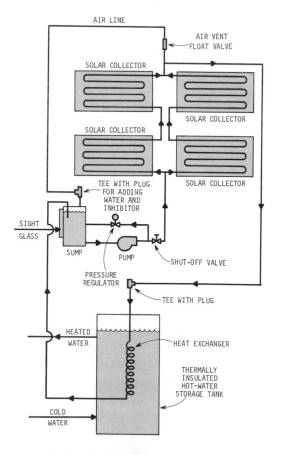

The Rho Sigma differential thermostat has a delay circuit that prevents excessive cycling of pump or blower. Here's how it works: The circuit is set so that the temperature difference between collector and storage must be greater to turn the pump or blower on than to turn it off. Without this feature, the thermostat could turn on the pump or blower, which would then send cool water or air from storage up to the collectors. This would cool the collectors, and the thermostat would shut the pump or blower off. The collectors would then warm up and again the thermostat would turn on the pump or blower, and the whole cycle would start again. Such excessive on-off cycling would be hard on the pump or blower, and hard on the contacts controlling the pump or blower.

RHO SIGMA DIFFERENTIAL THERMOSTAT

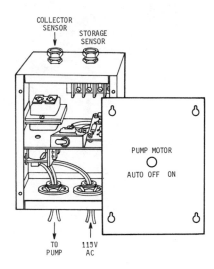

COLLECTOR SENSOR
STORAGE SENSOR
PUMP MOTOR
AUTO OFF ON
TO PUMP
115V AC

SAV WATER HEATER

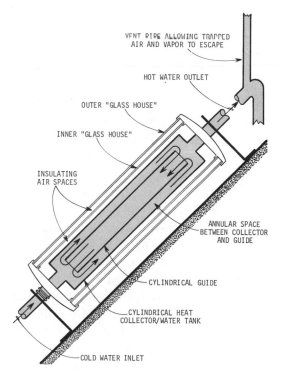

VENT PIPE ALLOWING TRAPPED AIR AND VAPOR TO ESCAPE

HOT WATER OUTLET

OUTER "GLASS HOUSE"

INNER "GLASS HOUSE"

INSULATING AIR SPACES

ANNULAR SPACE BETWEEN COLLECTOR AND GUIDE

CYLINDRICAL GUIDE

CYLINDRICAL HEAT COLLECTOR/WATER TANK

COLD WATER INLET

This SAV water heater combines absorber and storage in a single unit. A heavy, black-vinyl absorber/tank fits inside a double-walled glass cylinder. In operation, cold water enters the absorber and rises up the outer passage of the absorber/tank as it absorbs heat. Cooling water already near the top of the tank will travel downward through the core to be recirculated for further heating. As a result, a natural thermosiphon action keeps the water circulating through the tank, with the coolest water always being recycled for heating. This helps keep efficiency high.

Since this unit circulates potable water, and exposes it to outdoor conditions, it is most suitable in climates without freezing temperatures. At night the water will reverse cycle and begin to lose its heat. This prevents freeze-ups when outside temperatures drop as low as 10 degrees F., but it is a waste of stored heat. Overnight and cloudy-day storage is minimal with a unit like this, as compared to a system with an insulated storage tank inside the home.

A single SAV unit runs about $350 from Fred Rice Productions, 6313 Peach Ave., Van Nuys, CA 91401. Additional units can be used to increase capacity. Just hook them together end to end.

GRUMMAN SUNSTREAM WATER-HEATER SYSTEM

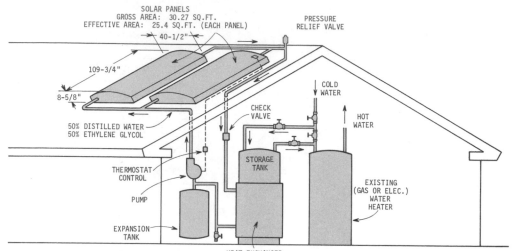

Grumman's Sunstream water-heater system features aluminum absorbers filled with a water/ethylene glycol solution. Collectors have curved cover plates to help dump snow loads, and they rinse clean in the rain. As with most solar-heating systems designed to interface with your existing conventional water heater, this unit circulates collection fluid via a thermostat-controlled pump. A heat exchanger—mounted outside the hot-water storage tank in this case—separates the toxic collection fluid from your water supply. Grumman says a system with two collector panels will supply about 50 gallons of hot water per sunny day. Price is around $1,000.

freeze-ups. You also need a check valve to keep pressurized water from backing up and filling the collectors as fast as they are drained. See the Reynolds Thorex system on the previous page.

Collector/storage System. All the heating systems mentioned so far feature separate collectors and storage tanks. But it is possible to combine collector and storage into one unit. A water

heater of this type might take the form of a blackened water tank inside a box glazed on its southern exposure and heavily insulated on all other surfaces. It might be a shallow blackened metal tray, also mounted in an insulated box with a glazed surface to let in the sun. Some commercially made water heaters such as the SAV consist of a cylindrical absorber/tank inside a two-walled glass cylinder. Hitachi makes a collector/tank by placing six black plastic cylinders in a glazed and insulated box.

Obviously such systems are not freeze-proof; the stored water is exposed to the weather at all times. If nighttime temperatures do not drop too far below freezing, however, some of these heaters will resist freeze-ups. The heat stored in the water is enough to keep the water from cooling all the way to the freezing point overnight. Of course this may reduce concern over freezing, but it is a waste of the heat collected during the day. For this reason, solar water heaters that combine collector and storage tank in a single unit are best suited to warm climates.

HOW TO BUY OR BUILD A SOLAR WATER HEATER

In the past few years, a number of commercially made solar water heaters have begun to appear in the U.S. market. Some are imported from Australia, Japan, or Israel—countries that led the U.S. in recognizing the need for rapid solar-energy development. Products made in the U.S. are mostly

SNELL UNDERGROUND SYSTEM

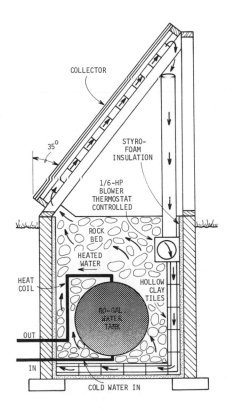

One way around the problem of freezing and collector leaks is to use an air collector to heat your water. Here's a system developed by Nebraskan John Snell. It consists of a freestanding collector built over an underground heat storage pit. Some 48 square feet of collector gathers heat and dumps it in the 180-cubic-foot pit filled with stones and an 80-gallon galvanized water tank. Air is moved by a 1/6-h.p. fan. It picks up hot air from the top edge of the collector and ducts it into the rock pit through the hollow cores of clay tiles lining the concrete-block walled pit. The heated air rises upward through the rocks, giving off heat, and is then circulated back through the collector.

As water leaves the tank to enter Snell's conventional water heater inside his home, it passes through a 40-foot coil of copper pipe placed in the rock bin. This coil is probably not necessary since the surface area of the tank exposed to hot rocks is already large enough for efficient heat exchange. The storage pit is lined with foam insulation. Pipe running to the house should be kept as short as possible, and should be insulated. Ordinary mineral wool won't do the job; it loses effectiveness when it gets wet. UMI Solar pipe insulation, available from Kalwall, is a good choice. Or you could gun plastic foam around the pipes.

The collector? Snell's design forces the air through aluminum beer cans with tops and bottoms removed, and painted black, of course. You can experiment with that design, or use the air collector shown earlier. You'll have to modify the design shown earlier by running the duct stubs out the bottom of the collector instead of out through the ends.

by firms in California and Florida. You can buy complete systems, have them installed by any competent plumber, and enjoy the benefits of solar hot water right now.

Or you can save money by making your own solar water heater. If you have the skills and the time, you can build most of your components from scratch. Or you can buy equipment such as collectors, tanks, and heat exchangers from various sources, and simply put them together. Obviously, you should be able to save money by doing most of the work yourself. And by working with high-quality materials, you should come up with a system that gives years of trouble-free service and a performance that matches that of any commercially made system. You'll have to decide on building or buying based on your assessment of your own financial resources and your manual skills. Take a look at the solar water heater plans, earlier in this chapter. If you think you can execute them, by all means build your own. If not, here's what you should know before buying.

Buying a Solar Water Heater. A solar water heater is a sizable investment. To pay off, it will have to provide trouble-free service for a number of years. Performance and durability are the two most important considerations. When you shop, think performance and durability first, price second. Here are some important steps to consider before buying any

solar water heater or solar space heater:

- Check out the firm that sells the unit. A quick call to your Better Business Bureau or chamber of commerce might save you years of grief. Before you buy, check over the sales contract and warranty very carefully. Better yet, let your lawyer do the job. Look for warranties of at least a year on mechanical components such as pumps and controls. You wouldn't be unreasonable to expect warranties for five years on simple structural parts such as collectors and plumbing.

- Examine the performance figures for the unit you plan to buy. If you are not sure, take your problem to a consulting engineer. Have him check the system out.

- Make sure the unit is approved by your local building codes. If it isn't, you may have to rip it out. Check your local tax situation, too. In some localities, energy-saving hardware is not subject to property taxes.

- Make sure you get a quotation on all costs, including installation. You don't want any surprises when the final bill comes in.

- Don't be unduly influenced by prices. Sure, it's smart to get the lowest price you can, but don't sacrifice quality and durability just to save money. Remember, a water heater is an investment. It adds to the value of your home, the same as any other home improvement. When you sell your home, you should be able to get your money back—if the system is still in good shape, still operating as it should.

With these guidelines in mind, what should you look for when buying a solar water heating system? First of all, favor copper construction. It will probably cost you more than an aluminum or plastic water heater, but it should hold up almost indefinitely. Copper is corrosion resistant. Although makers of aluminum heaters claim that corrosion is no problem when distilled or deionized water is used in conjunction with corrosion inhibitors, there is still room for doubt. Copper presents no such problem. It also has better heat transmission properties, so your collectors can be smaller if made of copper. Size can be important if your space available for collectors is limited. And smaller size will help offset copper's higher cost to some extent by reducing the costs of other collector materials such as glass and insulation.

What about plastic? So far, it seems to be practical. Special additives in the plastic have solved problems of deterioration due to ultraviolet light. Of course, corrosion is no problem. The drawbacks of plastic are these: First, it's not a great heat-transfer material, so your collectors must be oversized compared to copper or aluminum. Second, most plastic water heaters are designed to work best in warm climates, and in the underdeveloped countries of the world. Third, they are not suited to freezing climates in most cases. And fourth, they are difficult to tie into the typical American hot-water system.

Aluminum? Some large, reputable manufacturers are using the stuff, despite lingering fears of corrosion problems. Grumman, for example, is aggressively marketing its new Sunstream system which features aluminum collectors. Modern aluminum alloys are quite corrosion resistant, even in hard-use applications such as marine boat hulls and fittings. Still, the high temperatures present in any water-heating system tend to aggravate corrosion problems, so only time will tell the full story on the suitability of aluminum. If a maker is willing to stand behind his product with a good guarantee, you can probably rest assured that the product will hold up. If you can get a good guarantee from a reputable manufacturer, whom you expect will stay in business, by all means go ahead and try aluminum. I'd be less willing to place my confidence and money in the hands of a tiny newcomer to the field, someone without the experience required to avoid design and engineering errors, and without the reputation and stability that help ensure the manufacturer will still be in business 10 or 20 years from now.

SOLAR SPACE HEATING

Solar water heating may be the first logical step in your switch from fossil fuels to solar energy, but solar space heating is your biggest and most important step. If your home is typical, you probably use four times as much energy to heat your home as you use to heat water for consumption and washing. Obviously, then, the dollar benefits of switching to solar space heating can be much greater than those of switching to solar hot water.

Of course, the expenses for solar space heating are much greater. As it turns out, a solar space heating sys-

tem will probably save about four times as much energy as a basic water heating system, but it will also cost four times as much to install. Solar experts place the cost of a new home built with solar heating about $6,000 above the cost of a similar home without it. What do you get for that kind of money? Let's take a look at some of the systems being used today.

Water-cooled Collectors. First, because it is the most common form of space heating being used today, here's the basic heating system based on water-cooled collectors.

The heart of the system is the array

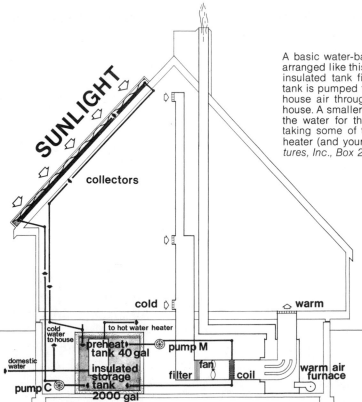

A basic water-based solar heating system is usually arranged like this. Heat from collectors is stored in an insulated tank filled with water. Hot water from the tank is pumped to a heat-exchange coil. A fan blows house air through the coil and distributes it to the house. A smaller tank of water inside the first preheats the water for the domestic hot water supply, thus taking some of the load off the conventional water heater (and your fuel bills). *Drawing by Acorn Structures, Inc., Box 250, Concord, MA 01742.*

Easiest way to get a solar house? Buy one complete from Acorn Structures. These feature water-cooled collector systems with copper piping and aluminum collector fins. The collectors drain whenever they are not collecting, so there's no danger of freeze-ups. Heat is stored in a 2,000-gallon tank which also holds a small preheating tank for the domestic hot water supply. The house is well designed with a variety of energy-efficient details including: maximum insulation, double glazing, minimal north window area, internal chimneys, and protective overhangs. Performance? A test house located in Acton, Massachusets, received about 46 percent of its space heating from the sun during the winter of 1975-1976. Prices of the home run from about $70,000 to $90,000 depending on the model you choose. The cost of these homes without the solar option would be $7,200 less. *Drawings courtesy Acorn Structures.*

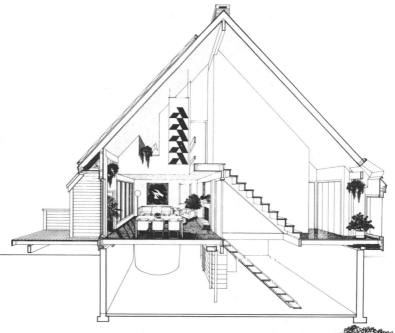

of solar collectors, as discussed earlier. These are placed on the roof in most cases, facing south, and tilted up from the horizontal at an angle equal to your latitude plus 10 to 15 degrees. That extra 10 or 15 degrees favors collection during the winter over summer collection, by pointing the collectors more directly at the low winter sun. The area covered by the collectors is roughly equal to half the floor space of the home. Obviously collector requirements will vary with the house and the climate. For an accurate method of determining your collector-area needs, see the section on sizing your solar installation, later in this chapter.

For the most reliability over the long haul, the collectors should be copper. For economy, they can be aluminum. If they can't go on the roof for one reason or another, they can be placed on the ground, or on a fence.

To pick up the heat gathered by the collectors, you pump a solution of antifreeze and water through the collectors via a system of CPVC plastic pipes. (These chlorinated-polyvinyl-chloride pipes withstand hot temperatures.) They lead down to the basement. There they enter the heat storage tank filled with some 1,500 gallons of water. The heat carried by the antifreeze solution is then transferred to the water in storage by means of a heat exchanger. Specifics: To circulate the heat transfer solution, the pump should be capable of circulating water at a rate of about 15 gallons per minute, and it should be

designed to handle hot water. The Grundfos pump sold by Kalwall is an efficient pump for the job. The heat exchanger will be located near the bottom of the tank so that the coolest water in storage will always receive the most heat. You can make your own exchanger by connecting 10 or 15 lengths of 3/4-inch copper pipe 10-feet long. Use U-bend fittings to connect the runs of pipe, laying them out in a serpentine pattern. Or simply use a 150-foot coil of 3/4-inch copper tubing.

To control the pump that drives this collection loop, you'll need a differential thermostat, as described earlier. The Deko Labs TC series or one of the Rho Sigma thermostats will do the job. To connect it, epoxy one sensor to the back of one of your collector panels, and surround it with insulation. Fasten the other to the pipe leading out of storage and back up to the collectors. Again, surround the sensor with insulation so that it is sensing only the temperature of the water in the pipe, not the air around it. Then the switching terminals hook up to the pump.

Storage tank? If you are starting with a new home, the best tank would be poured concrete, placed underneath the basement floor. If that's not possible, a steel tank inside the basement is the next choice. A tank outside the house is a poor choice since any heat it loses will be wasted. If you use a concrete tank, be sure it's made with a good rich mix of concrete. If you order from a ready-mix outfit, ask for a mix that conforms to ASTM specification C94. Basically, this calls for six bags of cement per yard, with a maximum of six gallons of water per bag of cement, plus other factors. This way your mix will not be too wet, and the concrete will be strong. It will be strong if it meets the C94 spec. A wet sloppy mix will be weak and prone to leaking. A good C94 mix should be waterproof. Even so, it's good practice to coat the inside of the tank with a waterproofing paint such as Thorseal. One of the new epoxy concrete sealers will do a good job, too, but cost is quite high.

Okay, your system can collect heat and carry it off for storage. Now it needs a means to extract that heat and put it to use warming your home. As noted before, you can't do this the easy way, by simply pumping the hot water from storage through hydronic baseboard heaters. The water is just not hot enough to allow good heat transfer from the baseboard unit to your room air. The solution is to pump the water through a finned-tube heat exchanger, then to blow air over the exchanger, and to route the heated air throughout your home through conventional warm air heating ducts. If you are adding a solar installation to a home already equipped with warm-air heat, the job will be easy. You can use the fan and ductwork already present in your current heating system. All you have to do is have a sheet-metal man place a finned-tube, liquid-to-air heat exchanger into the system. You can feed

water to this unit with a small pump. Pick this water up near the top of the storage tank where the water is hottest. Feed it through the exchanger and back to the bottom of the storage tank. Controls? A thermostat of the type used in conventional home heating applications does the job. When it calls for heat it should turn on both the pump feeding the liquid-to-air heat exchanger and the furnace fan as well.

Note: With this arrangement you can use the conventional warm-air furnace as a backup. Connect it to a separate house thermostat set a few degrees below the one controlling the solar heating system. Whenever the solar system can't do the job, the second thermostat will turn on the furnace. This second thermostat should be wired to shut off the first one whenever the second cuts in.

That's it for the basic hot-water solar space heating system. If you have the skills, you could put it together yourself. Or you could have the job done by one of the solar-heating contractors beginning to show up all across the country. In any case, the system is going to be a collection of components from various manufacturers. Solar heating requires such a broad mix of components that no one manufacturer builds complete systems. So your installation might include collectors from Revere, plastic piping from Genova, pumps by Grundfos, house thermostats by Honeywell, differential thermostats from Deko Labs, heat exchangers by York, and so on. Just remember that the system is only as strong as its weakest link. Use quality components at all points.

Variations? Instead of using a closed collection loop filled with antifreeze solution and equipped with a heat exchanger, you could use a drain-down system to protect against freezing. As I'll note later in the discussion of the Thomason Solaris system, there are advantages to this approach. If you decide to use the drain-down system, which drains the collectors whenever there is a chance of freezing, be sure to allow for proper drainage. If the system has low spots that will retain water, even a few ounces of the stuff can freeze up and cause trouble. See the drawing of the Reynolds system, earlier, for details.

You can use a conventional warm-air furnace as the backup for the solar system. This provides the ductwork and blower needed for the solar system. Another, cheaper alternative is to use a large water heater for backup heat. You'll still need ducts and a blower, plus a sheet metal plenum to house the liquid-to-air heat exchanger, but you won't have the expense of an entire furnace. The water heater provides backup heat whenever necessary by feeding its hot water through the heat exchanger in the hot-air distribution system. Again, a thermostat set a few degrees below the house thermostat controlling the solar system does the job of turning on the backup heater, and turning off the main thermostat.

A good woodburning stove makes a fine backup system, too. Its only disadvantage is that it won't be automatic; you have to be home to fire it up and keep it going. That limits your ability to leave home for any length of time during the heating season.

The Basic Air System. This is quite similar to the water system of space heating just described. But air takes the place of water. Ducts replace pipes. A bin full of rocks replaces the storage tank. Yet the principle remains the same.

An electric blower feeds air to the collectors through insulated ducts. This blower should be capable of supplying air at a rate around three cubic feet per minute, per square foot of collector. Thus, if you have 500 square feet of collector, the blower should be able to move about 1,500 cubic feet per minute. This is a common capacity for the typical $\frac{1}{2}$-horsepower heating and ventilating blower such as the Dayton K4C117. The blower has adjustable speeds so

A typical air, solar-heating system might be laid out as shown here. A system of motorized dampers lets the system collect heat, distribute heat directly from the collectors, or distribute heat from storage. As shown here, the system interfaces with a conventional hot-air furnace which supplies backup heat and the blower to run distribution from storage. If you eliminate the furnace in favor of a different backup system, move the collection blower to point A, downstream of the direct distribution return duct. In this position the blower can then move air in all three modes of operation (collection, distribution, and direct distribution).

TYPICAL AIR, SOLAR-HEATING SYSTEM

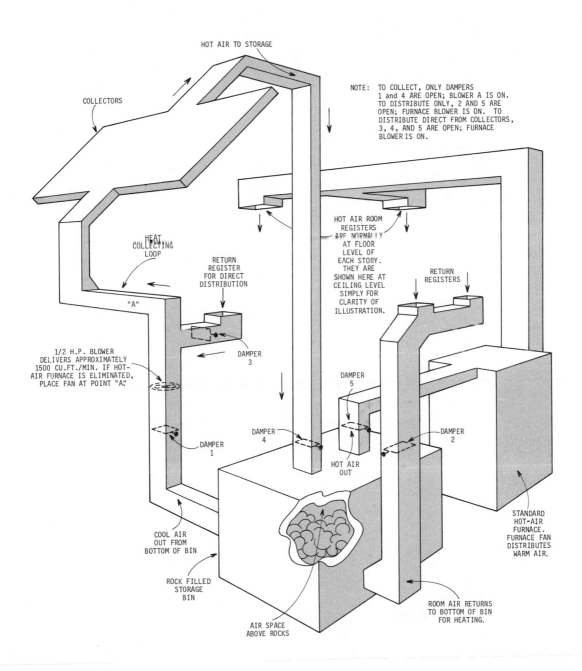

HOT AIR TO STORAGE

COLLECTORS

NOTE: TO COLLECT, ONLY DAMPERS
1 and 4 ARE OPEN; BLOWER A IS ON.
TO DISTRIBUTE ONLY, 2 AND 5 ARE
OPEN; FURNACE BLOWER IS ON. TO
DISTRIBUTE DIRECT FROM COLLECTORS,
3, 4, AND 5 ARE OPEN; FURNACE
BLOWER IS ON.

HEAT
COLLECTING
LOOP

"A"

RETURN
REGISTER
FOR DIRECT
DISTRIBUTION

HOT AIR ROOM
REGISTERS
ARE NORMALLY
AT FLOOR
LEVEL OF
EACH STORY.
THEY ARE
SHOWN HERE AT
CEILING LEVEL
SIMPLY FOR
CLARITY OF
ILLUSTRATION.

RETURN
REGISTERS

DAMPER
3

DAMPER
5

1/2 H.P. BLOWER
DELIVERS APPROXIMATELY
1500 CU.FT./MIN. IF HOT-
AIR FURNACE IS ELIMINATED,
PLACE FAN AT POINT "A."

DAMPER
4

HOT AIR
OUT

DAMPER
2

DAMPER
1

COOL AIR
OUT FROM
BOTTOM OF BIN

STANDARD
HOT-AIR
FURNACE.
FURNACE FAN
DISTRIBUTES
WARM AIR.

ROCK FILLED
STORAGE
BIN

ROOM AIR RETURNS
TO BOTTOM OF BIN
FOR HEATING.

AIR SPACE
ABOVE ROCKS

you can regulate flow for best operation.

The air blown through the collectors is ducted down in a closed loop to the basement where it passes through the storage bin full of rocks. From there the air is ducted back to the collectors for reheating. Hot air from the collectors should enter the storage bin at the top and exit at the bottom. Thus the coolest air in storage is always returned to the collectors to keep efficiency high.

This air flow is controlled by the usual differential thermostat. One probe of the thermostat is epoxied to the back of a collector plate; the other can be placed inside the storage bin. Or you can epoxy it to the duct leading out of storage back to the collectors. Both probes should be insulated to ensure they sense only the temperature of the collector or duct, not surrounding air. If your storage probe goes inside the bin, no insulation will be required. The same differential thermostat recommended for the water-based system can be used here—a Deko Labs model from the TC series, or a Rho Sigma, for example.

Extracting heat from storage is simply a matter of blowing house air through the bin of hot rocks. The huge surface area of the rocks inside the bin provides ample heat exchange. Note that the air flow for this extraction system should be the reverse of that for the collection air. That is, the air should enter the bin at the bottom, pass through the bin, and

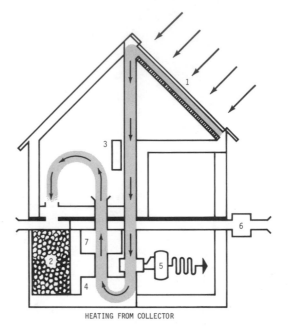

HEATING FROM COLLECTOR

Solaron offers a complete air system for space heating and for summertime cooling. The system can operate in five basic modes, heating your home directly from collectors or from storage. It can also store heat, store "cold," and cool from storage. System components include: 1) collectors; 2) rock storage bin; 3) control center that runs fans and dampers; 4) distribution and flow module containing dampers, ducts, blowers, filters; 5) preheating tank for hot water; 6) nighttime storage cooling unit; 7) auxiliary heater.

be drawn off for distribution from the top. A separate blower is usually used to move the air through this distribution network of insulated ducts.

One problem: When the fan blowing air through the collection loop is on, there's nothing to keep it from blowing air through the distribution network as well as the collectors. And when the distribution fan is going, there's nothing to keep it from blow-

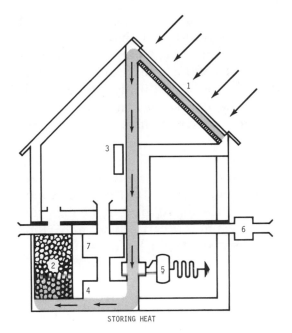

STORING HEAT

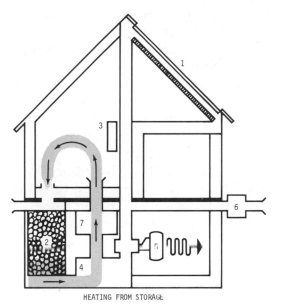

HEATING FROM STORAGE

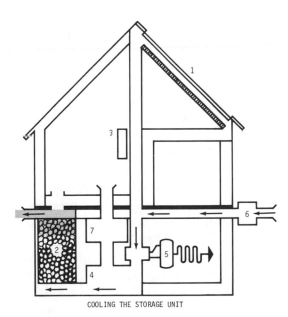

COOLING THE STORAGE UNIT

COOLING FROM STORAGE

ing air up through the collectors as well as through the distribution ducts. To keep air going to the right places, a system of automatic dampers is required.

Here's a common way to manage these dampers: When the house thermostat calls for heat, and the collectors are functioning, the dampers direct the hot air from the collectors directly through the distribution system. If the house thermostat calls for heat, but the collectors are not collecting (due to lack of sun), the dampers are set to allow air from storage to flow into the distribution system, but not into the collection loop. If the collectors are picking up heat, but the house thermostat is satisfied, the distribution system damper is closed, so that air being blown through the collection cycle will not detour through the distribution system. This type of damper system and its controls can get complicated. Installing it is best left to a heating contractor.

Backup system? A conventional warm air furnace is expensive, but its system of ducts and air distribution can double up to serve in your solar system. As in the water system described earlier, this furnace can have its own house thermostat, set a few degrees lower than the house thermostat controlling the solar system. Thus the backup system can switch on whenever the solar system fails to handle the load.

Solaris System. Harry Thomason's Solaris system is not yet available com-mercially, but you can order patented plans for building it yourself from Edmund Scientific Company, Barrington, NJ 08007. The system is one of the cheapest, and it has proven itself over the course of some 15 years in several working homes built by Thomason.

Collectors are made of corrugated aluminum roofing, insulated on their rear faces, covered with glass about two inches above the surface exposed to the sun. To carry away and store the heat captured by the collectors, Thomason pumps water from a basement storage tank up to the peak of the roof. From there it trickles down through the valleys in the corrugated roofing. The heated water flows into a trough and then through pipes leading back to the storage tank.

The tank is located in the basement, inside a large bin containing some 50 tons of rocks. These rocks pick up heat as it escapes from the tank of water, thus minimizing a common problem with water storage — its tendency to lose heat. The rocks capture this heat, and themselves become part of the storage system. They also serve as the heat exchanger for the Solaris system. To distribute heat from storage throughout the house, Thomason simply blows air through the rock bin. There it picks up heat and is then ducted throughout the house. Note the strong, cheap, simple method Thomason has developed for distributing incoming air evenly through the bin. This whole hybrid storage system is one example of

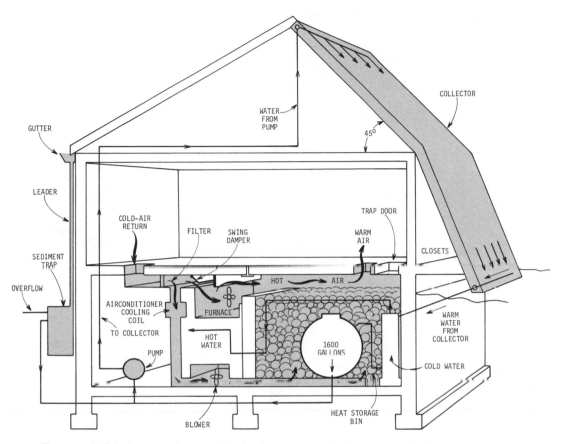

Thomason's Solaris system is one of the simplest, most practical, most tested solar heating systems around. It is also one of the cheapest to build. Plans and licenses for building it are available from Edmund Scientific, Edscorp Bldg., Barrington, N.J. 08007. See the accompanying text for a description of the system and its operation.

Thomason's brilliant efforts.

Thomason's approach to the problem of freezing is brilliant too. A differential thermostat controls the pump that sends water through the collectors. The thermostat works like any other in a solar application; it shuts off the pump when the collec-tors are cooler than the storage medium. Naturally the collectors are cooler than storage at night, or whenever there is a chance they might freeze. So the pump shuts off. The collectors drain down into storage, so there is no water left on the rooftop collectors to freeze. This eliminates

GARDEN WAY/SOL-R-TECH SYSTEM

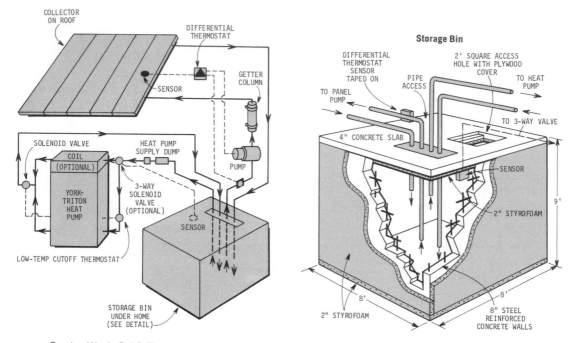

Garden Way's Sol-R-Tech system is unusual in that it uses a heat pump to extract heat from storage. This allows collectors to run at lower temperatures and higher efficiencies, and also lets the system get useful heat from fairly cool storage water. See the accompanying text for a full explanation of the system and a discussion of its built-in advantages and disadvantages.

the need for antifreeze and heat exchangers. It also eliminates the need to heat up ice-cold antifreeze every morning, a waste of solar heat inherent in any system that features collectors that remain filled with water at all times.

Garden Way/Sol-R-Tech. This solar heating system starts out like most other water-cooled systems. It features aluminum collectors, a water storage medium housed in an un-

derhouse tank, and a differential-thermostat-controlled pump to circulate the fluid through the collectors.

The big difference between the Garden Way system and most others is that Garden Way advocates the use of a heat pump to extract heat from storage and to distribute it through a system of ducts. This approach has good points and bad. On the plus side: The heat pump can extract heat from storage, even when the water in storage has dropped as low as 40 degrees. This lets the collectors run at low temperatures and high efficiencies. (As noted before, collectors are most efficient when run as cool as possible.) Garden Way claims the use of the heat pump doubles the efficiency of its collectors. Another advantage is that the heat pump will run in reverse during the summer to provide airconditioning with no extra investment in equipment.

Bad points? The heat pump adds to the cost of the installation. And the heat pump requires more power than would be needed to extract heat by blowing air across the coils of a finned-tube heat exchanger. Garden Way's decision to recommend the heat pump is largely influenced by the Vermont climate where the Garden Way labs are located. The pump makes better use of the limited amount of sunshine available there. In a warmer and sunnier climate, a conventional heat exchanger would out-perform a heat pump, and would cost less to install and operate.

Garden Way realizes this, and they recommend an optional heat exchanger be incorporated into the system. Whenever weather is favorable and the water in storage can be raised to the 100-degree point, a solenoid valve diverts storage water from the heat pump to the heat exchanger. At the same time, it turns off the heat pump and activates the heat-exchanger blower. This allows the system to provide heat directly from storage whenever the water is hot enough to work effectively through a heat exchanger. As a result, the extra power is saved that would otherwise be running the heat pump instead of the heat-exchanger blower. As soon as the water temperature drops below 100 degrees, however, the heat exchanger will no longer be able to extract heat effectively. So the solenoid valve switches the storage water back to the heat pump, which resumes its job of extracting the heat and distributing it throughout the house.

The solenoid valve is controlled by a thermostat with its sensor inside the storage tank. The thermostat is set to switch at 100 degrees.

This rather elaborate system worked out by Garden Way requires one other set of controls. This is a switch to turn off the heat pump whenever the water in storage drops below 40 degrees. When the water gets below 40 the heat pump compressor has to work extra hard to extract heat. Damage may result. There is also the danger of freezing in the event the pump extracts too much heat from water below 40 degrees. Again, damage may result. To prevent this, Garden Way recommends the installation of its Klixon thermostat. Tape the thermostat to the pipe feeding water from storage to the heat pump. Wire it in series with the heat pump's compressor contactor.

Warning: As noted in Garden Way's *Sol-R-Tech Operations Manual*, the solar heating system has no protection against freeze-ups. The manual recommends circulating storage water directly through the collectors and outlines precautions to take to prevent corrosion (using softened water and a corrosion inhibitor, plus a getter column), but it neglects the freezing problem entirely. Best solution is to run distilled or softened water, plus automotive antifreeze (for aluminum block engines) through the collectors and a heat exchanger submerged in the storage tank. Your other choice would be to fill the storage tank and collectors with a solution of water and antifreeze. This would require almost 2,000 gallons of antifreeze, given the size of the storage tank Garden Way recommends.

RETROFITTING AN EXISTING HOME

Sure, the idea of heating your home with free energy from the sun has great appeal. And sure, the best way to get an efficient solar home is to build it from scratch. But most people don't want to move out of their existing homes and into new custom-built solar homes just so they can enjoy the benefits of solar heat.

So for most people, the big question about solar heating is this: Can my home be heated by the sun? The answer is, that it probably can, but if you want the job to perform economically, you'll probably have to make some fairly major improvements on your home first. You may even be forced to change your landscaping. Here are some of the conditions

you'll have to meet before you can expect economical performance from any solar heating system:

• Obviously, you'll need someplace to place your collector array. This should be an area open to direct sunlight all day long during the winter months. The collectors should face directly south, or as much as 15 degrees to the west of south. They should be tipped back at the top so that they form an angle with the horizontal equal to the latitude of your site plus 10 to 15 degrees. Thus, if you live at 40 degrees north latitude, your collectors should slope at a 50- to 55-degree angle.

Mounting your collectors at an angle like this places their faces perpendicular to the sun's rays. This lets the collectors absorb the maximum amount of radiation possible, while cutting down on reflective losses. The roof is the most popular location for collectors. This puts them up out of the way where they are less susceptible to damage than they would be on the ground. Trouble is, you can do this only if your roof ridgeline runs east and west, or close to it. And the pitch of most roofs is less than the ideal pitch for the collectors.

Actually, there is room for compromise in orienting collectors, so if your home doesn't fit the requirements listed above, don't despair. University of Pennsylvania tests have shown that when your collec-

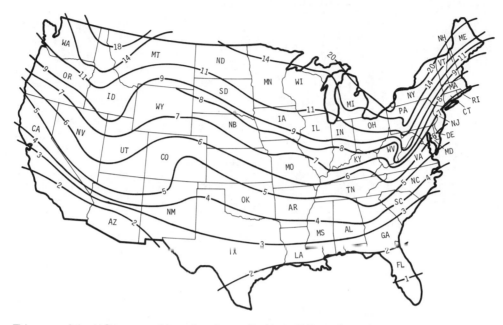

This map of the U.S., prepared by solar pioneer Dr. Maria Telkes, shows the relative practicality of solar energy in all parts of the country. The map is calibrated in degree days per sunshine hour, based on data collected during the months of January and February. The higher the number, the more difficult it becomes to heat with solar energy. With a large enough solar collector area, however, solar heating is possible anywhere on this map — if basic cost is no factor.

tors face 23 degrees east or west of due south they lose only about five percent of their efficiency. Increase this to 33 degrees and you lose about 10 percent. As for tilt, you can go as far as 30 degrees either side of the ideal angle without much loss.

So you can place your collectors in less than ideal position without suffering great losses in efficiency. Still, it's unwise to do so unless you have no other choice. If you lose 10 percent of your heating capability due to collector location or slant, the only way to get the heat back is to increase your collector area by 10 percent. And that increases your costs.

While placing collectors on the roof is the most common practice, you can place your array on the ground, or build it into a fence. If you decide on either of these approaches, you'll want to put your collectors where they will be least conspicuous, and least susceptible to damage. And of course you'll want to place them where they are exposed to the sun.

• If you have a place to put your collectors, your next consideration

(Text continued on page 143)

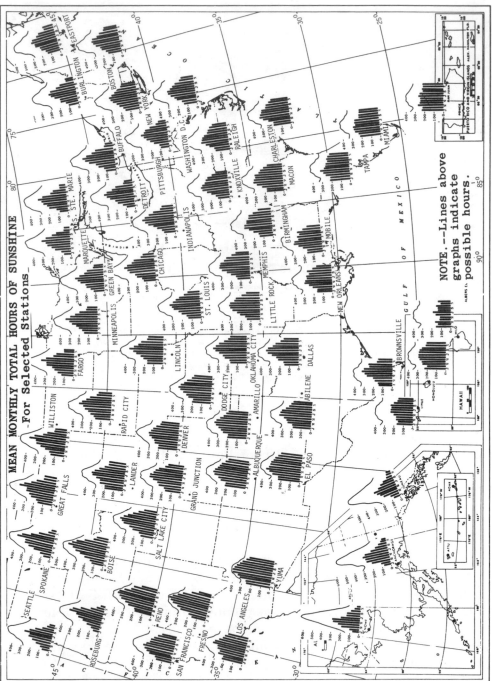

MEAN MONTHLY TOTAL HOURS OF SUNSHINE
—For Selected Stations—

NOTE.—Lines above graphs indicate possible hours.

This map and the table and map on following pages can help you evaluate the solar potential of your present home, and guide you to solar hotspots around the country should you ever decide to move. If you are interested in solar

energy primarily for space heating, the data for the months of the heating season are most important. If you plan on solar water heating, the data for every month are important. Note that the figures given here are averages made over

a 30-year period in most cases. They are general figures only, and may not apply to specific sites within the various regions. These maps and the table are reprinted from *The Climatic Atlas of the United States*.

MEAN NUMBER OF HOURS OF SUNSHINE

STATE AND STATION	YEARS	JAN.	FEB.	MAR.	APR.	MAY	JUNE	JULY	AUG.	SEPT.	OCT.	NOV.	DEC.	ANNUAL
ALA. BIRMINGHAM	30	138	152	207	248	293	294	269	265	244	234	182	136	2662
MOBILE	22	157	158	212	253	301	289	249	259	235	254	195	146	2708
MONTGOMERY	30	160	168	227	267	317	311	288	290	260	250	200	156	2894
ALASKA ANCHORAGE	19	78	114	210	254	268	288	255	184	128	96	68	49	1992
FAIRBANKS	20	54	120	224	302	319	334	274	164	122	85	71	36	2105
JUNEAU	29	71	102	171	200	230	251	193	161	123	67	60	51	1680
NOME	27	72	109	193	226	285	297	204	146	142	101	67	42	1884
ARIZ. PHOENIX	30	248	244	314	346	404	404	377	351	334	307	267	236	3832
PRESCOTT	14	222	230	293	323	378	392	323	305	315	286	254	228	3549
TUCSON	13	255	266	317	350	399	394	329	329	335	317	280	258	3829
YUMA	30	258	266	337	365	419	420	404	380	351	330	285	262	4077
ARK. FT. SMITH	30	146	156	202	234	268	303	321	305	261	230	174	147	2747
LITTLE ROCK	30	143	158	213	243	291	316	321	316	265	251	181	142	2840
CALIF. EUREKA	30	120	138	180	209	247	261	244	205	195	164	127	108	2198
FRESNO	29	153	192	283	330	389	418	435	406	355	306	221	144	3632
LOS ANGELES	30	224	217	273	264	292	299	352	336	295	263	249	220	3284
RED BLUFF	15	156	186	246	302	366	396	438	407	341	277	199	154	3468
SACRAMENTO	30	134	169	255	300	367	405	437	406	347	283	197	122	3422
SAN DIEGO	30	216	212	262	242	261	253	293	277	255	234	236	217	2958
SAN FRANCISCO	30	165	182	251	281	314	330	300	272	267	243	198	156	2959
COLO. DENVER	30	207	205	247	252	281	311	321	297	274	246	200	192	3033
GRAND JUNCTION	30	169	182	243	265	314	350	349	311	291	255	198	168	3095
PUEBLO	30	224	217	261	271	299	340	349	318	290	265	225	211	3270
CONN. HARTFORD	30	141	166	206	223	267	285	299	268	220	193	137	136	2541
NEW HAVEN	30	155	178	215	234	274	291	309	284	238	215	157	154	2704
D. C. WASHINGTON	30	138	160	205	226	267	288	291	264	233	207	162	135	2576
FLA. APALACHICOLA	26	193	195	233	274	328	296	273	259	236	263	216	175	2941
JACKSONVILLE	30	192	189	241	267	296	260	255	248	199	205	191	170	2713
KEY WEST	30	229	238	285	296	307	273	277	269	236	237	226	225	3098
LAKELAND	7	204	186	222	251	285	268	252	242	203	209	212	198	2732
MIAMI	30	222	227	266	275	280	251	267	263	216	215	212	209	2903
PENSACOLA	30	175	180	232	270	311	302	278	284	249	265	206	166	2918
TAMPA	30	223	220	260	283	320	275	257	252	232	243	227	209	3001
GA. ATLANTA	25	154	165	218	266	309	304	284	285	247	241	188	160	2821
MACON	30	177	178	235	279	321	314	292	295	253	236	202	168	2950
SAVANNAH	30	175	173	229	274	307	279	267	256	212	216	197	167	2752
HAWAII HILO	7	153	135	161	112	106	158	184	134	137	153	106	131	1670
HONOLULU	30	227	202	250	255	276	280	293	290	279	257	221	211	3041
LIHUE	10	171	162	176	176	211	246	246	236	246	210	170	161	2411
IDAHO BOISE	30	116	144	218	274	322	352	412	378	311	232	143	104	3006
POCATELLO	30	111	143	211	255	300	338	380	347	296	230	145	108	2864
ILL. CAIRO	15	124	160	218	254	298	324	345	336	279	254	181	145	2918
CHICAGO	30	126	142	199	221	274	300	333	299	247	216	136	118	2611
MOLINE	18	132	139	189	214	255	279	337	300	251	214	130	123	2563
PEORIA	30	134	149	198	229	273	303	336	299	259	222	149	122	2673
SPRINGFIELD	30	127	149	193	224	282	304	346	312	266	225	152	122	2702
IND. EVANSVILLE	30	123	145	199	237	294	322	342	318	274	236	156	120	2766
FT. WAYNE	30	113	136	191	217	281	310	342	306	242	210	120	102	2570
INDIANAPOLIS	30	118	140	193	227	278	313	342	313	265	222	139	118	2668
TERRE HAUTE	24	125	148	189	231	274	302	341	305	253	235	150	122	2675
IOWA BURLINGTON	19	148	165	217	241	284	315	353	327	270	243	175	147	2885
CHARLES CITY	22	137	157	190	226	258	285	336	290	241	207	130	115	2572
DES MOINES	30	155	170	203	236	276	303	346	299	263	227	156	136	2770
SIOUX CITY	30	164	177	216	254	300	320	363	320	270	236	160	146	2926
KAN. CONCORDIA	30	180	172	214	243	281	315	348	308	249	245	189	172	2916
DODGE CITY	30	205	191	249	268	305	335	359	335	290	266	218	198	3219
TOPEKA	18	159	160	193	215	260	287	310	304	263	229	173	149	2702
WICHITA	30	187	186	233	254	291	321	350	325	277	245	206	182	3057
KY. LOUISVILLE	30	115	135	188	221	283	303	324	295	256	219	148	114	2601
LA. NEW ORLEANS	30	160	158	213	247	292	287	260	269	241	260	200	157	2744
SHREVEPORT	19	151	172	214	240	298	332	339	322	289	273	208	177	3015
MAINE EASTPORT	22	133	151	196	201	245	248	275	260	205	175	105	115	2309
PORTLAND	30	155	174	213	226	268	286	312	294	229	202	146	148	2653
MD. BALTIMORE	30	148	170	211	229	270	295	299	272	238	212	164	145	2653
MASS. BLUE HILL OBS.	10	125	136	165	182	233	248	266	241	211	181	134	135	2257
BOSTON	30	148	168	212	222	263	283	300	280	232	207	152	148	2615
NANTUCKET	22	128	156	214	227	278	284	291	279	242	208	149	129	2585
MICH. ALPENA	24	86	124	198	228	261	303	339	285	204	159	70	67	2324
DETROIT	30	90	128	180	212	263	295	321	284	226	189	98	89	2375
LANSING	30	84	119	175	215	272	305	344	294	228	182	87	73	2378
ESCANABA	30	112	148	204	226	266	283	316	267	198	162	90	94	2366
GRAND RAPIDS	30	74	117	178	218	277	308	349	304	231	188	92	70	2406
MARQUETTE	30	78	113	172	207	248	268	305	251	186	142	68	66	2104
SAULT STE. MARIE	30	83	123	187	217	252	269	309	256	165	133	61	62	2117
MINN. DULUTH	30	125	163	221	235	268	282	328	277	203	166	100	107	2475
MINNEAPOLIS	30	140	166	200	231	272	302	343	296	237	193	115	112	2607
MISS. JACKSON	12	130	147	199	244	280	287	279	287	235	223	185	150	2646
VICKSBURG	30	136	141	199	232	284	304	291	297	254	244	183	140	2705
MO. COLUMBIA	30	147	164	207	232	281	296	341	298	262	225	166	138	2757
KANSAS CITY	30	154	170	211	235	278	313	347	308	266	235	178	151	2846
ST. JOSEPH	23	154	165	211	231	274	301	347	287	260	224	168	144	2766
ST. LOUIS	30	137	152	202	235	283	301	325	289	256	223	166	125	2694
SPRINGFIELD	30	145	164	213	238	278	305	342	310	269	233	183	140	2820
MONT. BILLINGS	21	140	154	208	236	283	301	372	332	258	213	136	129	2762
GREAT FALLS	19	154	176	245	261	299	299	381	342	256	206	132	133	2884
HAVRE	30	136	174	234	268	311	312	384	339	260	202	132	122	2874
HELENA	30	138	168	215	241	292	292	342	336	258	202	137	121	2742
MISSOULA	25	85	109	167	209	261	260	378	328	246	178	90	66	2377

STATE AND STATION	YEARS	JAN.	FEB.	MAR.	APR.	MAY	JUNE	JULY	AUG.	SEPT.	OCT.	NOV.	DEC.	ANNUAL
NEBR. LINCOLN	30	173	172	213	244	287	316	356	309	266	237	174	160	2907
NORTH PLATTE	30	181	179	221	246	282	310	343	304	264	242	184	169	2925
OMAHA	30	172	188	222	259	305	332	379	311	270	248	166	145	2997
VALENTINE	30	185	194	229	252	296	323	369	326	275	242	174	172	3037
NEV. ELY	22	186	197	262	260	300	354	359	344	303	255	204	187	3211
LAS VEGAS	8	239	251	314	336	386	411	383	364	345	301	258	250	3838
RENO	30	185	199	267	306	354	376	414	391	336	273	212	170	3483
WINNEMUCCA	30	142	155	207	255	312	346	395	375	316	242	177	139	3061
N. H. CONCORD	23	136	153	192	196	229	261	286	260	214	179	122	126	2354
MT. WASHINGTON OBS.	18	94	98	133	141	162	145	150	143	139	159	89	87	1540
N. J. ATLANTIC CITY	30	151	173	210	233	273	287	298	271	239	218	177	153	2683
TRENTON	30	145	168	203	235	277	294	309	273	239	208	160	142	2653
N. MEX. ALBUQUERQUE	30	221	218	273	299	343	365	340	317	299	279	245	219	3418
ROSWELL	21	218	223	286	306	330	333	341	313	266	266	242	216	3340
N. Y. ALBANY	30	125	151	194	213	266	301	317	286	224	192	115	112	2496
BINGHAMTON	30	94	119	151	170	226	256	266	230	184	158	92	79	2025
BUFFALO	30	110	125	180	212	274	319	338	297	239	183	97	84	2458
NEW YORK	30	154	171	213	237	268	289	302	271	235	213	169	155	2677
ROCHESTER	30	93	123	172	209	274	314	333	294	224	173	97	86	2392
SYRACUSE	30	87	115	165	197	261	295	316	276	211	163	81	74	2241
N. C. ASHEVILLE	30	146	161	211	247	289	292	268	250	235	222	179	146	2646
CAPE HATTERAS	9	152	168	206	259	293	301	286	265	214	202	169	154	2669
CHARLOTTE	30	165	177	230	267	313	316	291	277	247	243	198	167	2891
GREENSBORO	30	157	171	217	231	298	302	287	272	243	236	190	163	2767
RALEIGH	29	154	168	220	255	290	284	277	253	224	215	184	156	2680
WILMINGTON	30	179	180	237	279	314	312	286	273	237	238	206	178	2919
N. DAK. BISMARCK	30	141	170	205	236	279	294	358	307	243	198	130	125	2686
DEVILS LAKE	30	150	177	220	250	291	297	352	302	230	198	123	124	2714
FARGO	30	132	170	210	232	283	288	343	293	222	187	112	114	2586
WILLISTON	29	141	168	215	260	305	312	377	328	247	206	131	129	2819
OHIO CINCINNATI (ABBE)	30	115	137	186	222	273	309	323	295	253	205	138	118	2574
CLEVELAND	30	79	111	167	209	274	301	325	288	235	187	99	77	2352
COLUMBUS	30	112	132	177	215	270	296	323	291	250	210	131	101	2508
DAYTON	30	114	136	195	222	281	313	323	307	268	229	152	124	2664
SANDUSKY	30	100	128	183	229	285	312	343	302	248	201	111	91	2533
TOLEDO	30	93	120	170	203	263	296	331	298	241	196	106	92	2409
OKLA. OKLAHOMA CITY	29	175	182	235	253	290	329	352	331	282	243	201	175	3048
TULSA	[illegible]	152	164	200	213	244	287	314	308	281	241	207	172	2783
OREG. BAKER	22	118	143	198	251	302	313	406	368	289	215	132	100	2835
PORTLAND	30	77	97	142	203	246	249	329	275	218	134	87	65	2122
ROSEBURG	30	69	96	148	205	257	278	369	329	255	146	81	50	2283
PA. HARRISBURG	30	132	160	203	230	277	297	319	282	233	200	140	131	2604
PHILADELPHIA	30	142	166	203	231	270	281	288	253	225	205	158	142	2564
PITTSBURGH	[illegible]	89	114	163	200	239	260	283	250	234	180	114	76	2202
READING	30	133	151	195	220	259	275	293	259	219	198	144	127	2473
SCRANTON	30	108	138	178	199	251	269	290	249	213	183	120	105	2303
R. I. PROVIDENCE	30	145	168	211	221	271	285	292	267	226	207	153	143	2589
S. C. CHARLESTON	30	188	189	243	284	323	308	297	281	244	239	210	187	2993
COLUMBIA	30	173	183	233	274	312	312	291	283	243	242	202	166	2914
GREENVILLE	[illegible]	166	176	227	274	307	300	278	274	239	232	192	157	2822
S. DAK. HURON	30	153	177	213	250	295	321	367	320	260	212	142	134	2844
RAPID CITY	30	164	182	222	245	278	300	348	317	266	228	164	144	2858
TENN. CHATTANOOGA	30	126	146	187	239	290	295	278	266	247	220	169	128	2591
KNOXVILLE	30	124	144	189	237	281	288	277	248	237	213	157	120	2515
MEMPHIS	30	135	152	204	244	296	321	319	314	261	243	180	139	2808
NASHVILLE	30	123	142	196	241	285	308	292	279	250	224	168	126	2634
TEX. ABILENE	[illegible]	190	199	250	259	290	347	335	322	276	245	223	201	3137
AMARILLO	30	207	199	258	276	305	338	350	328	288	260	229	205	3243
AUSTIN	30	148	152	207	221	266	302	331	320	261	242	180	160	2790
BROWNSVILLE	30	147	152	187	210	272	297	326	311	246	252	165	151	2716
CORPUS CHRISTI	24	160	165	212	237	295	329	366	341	276	264	194	164	3003
DALLAS	30	155	159	220	238	279	326	341	325	274	240	191	163	2911
DEL RIO	27	173	173	230	237	259	279	331	319	252	240	195	178	2866
EL PASO	30	234	236	299	329	373	369	336	327	300	287	257	236	3583
GALVESTON	30	151	149	203	230	288	322	305	292	257	264	199	151	2811
HOUSTON	30	144	141	193	212	266	298	294	281	238	239	181	146	2633
PORT ARTHUR	30	153	149	209	235	292	317	285	281	252	256	191	148	2768
SAN ANTONIO	30	148	153	213	224	258	292	325	307	261	241	183	160	2765
UTAH SALT LAKE CITY	30	137	155	227	269	329	358	377	346	306	249	171	135	3059
VT. BURLINGTON	30	103	127	184	185	244	270	291	266	199	152	77	80	2178
VA. LYNCHBURG	26	153	169	216	243	288	297	288	264	235	217	177	158	2705
NORFOLK	30	156	174	223	257	304	311	296	282	237	220	182	161	2803
RICHMOND	30	144	166	211	248	280	296	286	263	230	211	176	152	2663
WASH. NORTH HEAD	22	76	97	135	182	221	214	226	186	170	123	87	66	1783
SEATTLE	30	74	99	154	201	247	234	304	248	197	122	77	62	2019
SPOKANE	30	78	120	197	262	308	309	397	350	264	177	86	57	2605
TATOOSH ISLAND	30	70	100	135	182	229	217	235	190	175	129	71	60	1793
WALLA WALLA	30	72	106	194	262	317	335	411	367	280	198	92	51	2685
W. VA. ELKINS	24	110	119	158	198	227	256	225	236	211	186	131	103	2160
PARKERSBURG	30	91	111	155	200	252	277	286	264	230	189	117	93	2265
WIS. GREEN BAY	30	121	148	194	210	251	279	314	266	213	176	110	106	2388
MADISON	30	126	147	196	214	258	285	336	288	230	198	116	108	2502
MILWAUKEE	30	116	134	191	218	267	293	340	292	235	193	125	106	2510
WYO. CHEYENNE	30	191	197	243	237	259	304	318	286	265	242	188	170	2900
LANDER	30	200	208	260	264	301	340	361	326	280	233	186	185	3144
SHERIDAN	30	160	179	226	245	286	303	367	333	266	221	153	145	2884
P. R. SAN JUAN	30	231	229	273	252	240	245	264	257	219	229	217	222	2878

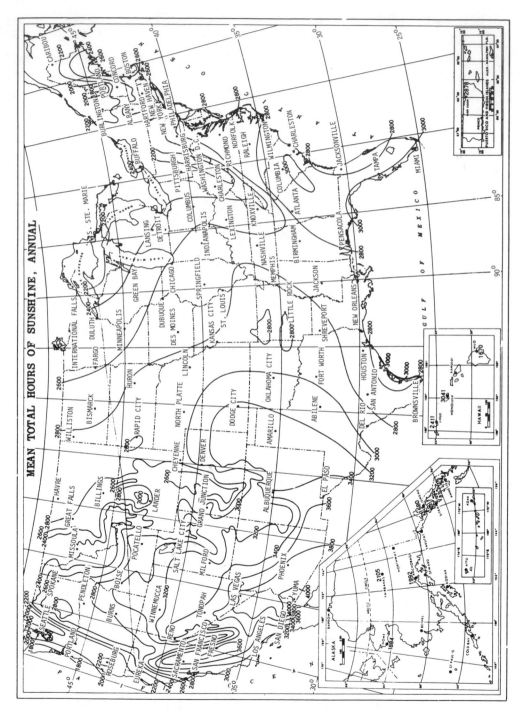

MEAN TOTAL HOURS OF SUNSHINE, ANNUAL

is to make sure your home is energy-efficient. It makes no sense to waste money on a solar heating system hooked onto a poorly-insulated, drafty home. You'll save more energy and more money by improving the home than you will by installing a solar heating system. Before you even think about installing solar heat, be sure to look over the chapters on insulation, caulking and weatherstripping, windows and doors, and energy-efficient architecture. Do everything you can to make your home an energy saver. The more you do, the better your solar heating system can serve you. A system that provides only half the heating needs of an energy-wasting home could take over the full load if that home is brought up to modern energy saving specifications.

- Adequate sunlight is the final — and the most important — requirement you'll have to meet. If you live in an area of the country where there is very little sunshine, solar heat won't pay. Fortunately, there are very few parts of the U.S. where there is not enough sunshine to make solar heating practical. If you live in a chronically cloudy area, you'll discover that a solar installation is possible, but it will require so much collector area that costs may rule out installation. See the sizing section coming up.

And don't make the mistake of thinking you can't use solar energy just because you live in a cold climate. There's still plenty of energy available up north, as long as the sun shines. The temperature of your climate is not as important as the number of bright, sunny days.

- One last note: In addition to meeting the above requirements, you'll have to provide space for heat storage. Best place is inside the house, in the basement. Thus any heat that manages to leak out of storage will remain in the house where it will still contribute heat. If you have no basement you'll do best burying your storage under ground. Either way, figure that a rock storage bin for an air system will take up the space of a cube eight feet on a side. Water storage will require a tank of about one-third that volume, about 1,500 gallons.

Sizing Your Solar Heating System. This shouldn't be very difficult, if you will be retrofitting your home, and if you have been keeping records on your fuel consumption over past years. If you are designing a system for a new home, or if you don't have figures for your old home, things will get a bit tricky. In either case, your goal is the same: Balance the heating requirements of your home against the output of your collectors.

To determine the heating load of your present home, check your fuel consumption records from past years. If you don't have them, your fuel company may be able to help you.

Once you have the records, the rest

is easy. Take a look at the fuel-value table below. Then multiply the value given below (for the fuel you use) by the amount of fuel you use per year. The result is your heating requirement, in Btu per heating season. The chart assumes 70-percent efficiencies for gas and oil furnaces, and 100-percent efficiencies for electric heat.

HOME HEATING FUEL VALUES

Fuel Oil	91,000 Btu per gallon
Gas	700 Btu per cubic foot
Electricity	3,412 Btu per Kwhr

Once you know your Btu load for the season, simply match it against the ability of the collectors you have chosen. Most collector makers can supply you with data on their collectors, telling how many Btu they will deliver in your area. These figures are often given in terms of Btu per each month of the year. So simply add up the Btu the collectors can deliver during those months when your home requires heat. If you match Btu collected against Btu required, you'll come up with a system that provides about 70 percent of your heating needs, not 100 percent. Why? Because you'll probably have a 30-percent loss in your distribution and storage systems.

Today, most experts recommend a system of about 70-percent capacity because it will pay off soonest when balanced against today's costs for heating fuels. Of course prices as this is written won't hold for long, so you might consider expanding your collector area to provide closer to 100 percent of your needs. Remember that as the price of heating fuel rises, the economic practicality of 100-percent solar heating rises. On the other hand, you can always add to your collector area at a later date as increasing fuel prices make the addition cost-effective.

If you don't have records of heating fuel consumption for past years, you'll have to run a heat loss calculation. This can be fairly complicated, but one relatively simple technique has been worked out by the National Association of Home Builders. The technique is outlined in their *Insulation Manual for Homes and Apartments*. It's $4 from NAHB Research Foundation, 627 Southlawn Lane, Box 1627, Rockville, MD 20850. This book is also a good source of information on insulating your home. But if you'd prefer not to tackle the calculations, a heating engineer can make a heat-loss evaluation of your home.

You might go through all these calculations and find that you'll need a very large and expensive collector area. This tells you that you live in a poor climate for solar heating—assuming your house is reasonably energy efficient. In this case, you might consider other forms of home heating.

Most areas of the country where solar heating is most expensive are also areas where there is a large amount of firewood available at reasonable cost. Here, a woodstove or two might be a better answer than solar heat. I provide about two-thirds of my heating requirements with a small woodstove, and the whole installation cost a tiny fraction of what I'd have had to pay for a solar system. Of course the cutting, splitting, stacking and stoking require a lot more effort than collecting sunshine. But I look at it as healthy exercise. If the idea of wood heat appeals to you, check out Chapter 10.

Sizing Your Storage Unit. This is easy. For residential purposes, a tank of water with a capacity of 1,500 or 2,000 gallons will do the job—if you choose water as your storage medium. If you choose rock, 15 or 20 tons of rock will do the job. A ton of gravel is roughly equal to a cubic yard, and a cubic yard is 27 cubic feet. Therefore, 20 tons of rock will occupy about 540 cubic feet, or a bin measuring about 8 feet high, 12 feet long and 8 feet wide. Or this bin could be a cube 8 feet on side.

It's not important to match storage area exactly to your home or your collector area. The capacity of any storage system in this size range will exceed the collection capabilities of all but huge areas of collector. Theoretically, water and rock will stop absorbing heat for storage purposes only when they reach boiling temperatures.

SIMPLE SOLAR HEATING SYSTEMS

America's infatuation with high technology often results in a tendency to oversolve its problems. Nobody knows this better than a couple of brilliant solar pioneers, Harold Hay and Steve Baer. Both of these energy prophets have come up with passive solar heating systems that are so simple nobody will believe they work. Yet work they do, in some cases better than any other system available. Let's take a look.

Harold Hay's Sky-Therm Principle. This is so simple and so elegant, yet so effective, it should be mandatory in all new homes built in the Southwest. It's a total concept, a combination of house design and climate control system.

The basic home, as shown in the drawings on page 147, is a simple one-story structure with concrete block walls and partitions. The cores of the partition wall blocks are filled with sand to increase their mass, and thus their ability to store heat. Most of the cores in the blocks in the exterior walls are filled with vermiculite to increase the insulation value.

The roof is flat, made of corrugated steel panels. Over it goes a sheet of black vinyl plastic. Then, covering the entire roof of the home, are four large clear vinyl "waterbeds" about nine inches deep. This puts a total of some 6,300 gallons of water (26 tons) on the roof. Suspended above these waterbeds is a sheet of UV-resistant vinyl, transparent to let the sun shine through.

Between the "waterbeds" are aluminum tracks running north and south. On these tracks ride 2-inch-thick panels of urethane-foam insulation. An electric motor can slide these panels into position to completely cover the waterbeds, or into a storage position on the roof of the home's open carport.

Okay, that's how the home is built. Here's how it works, automatically providing its own heat in winter, its own cooling during the summer.

Winter heating. During the daytime, the movable foam insulation panels are automatically rolled back into their storage position on the carport roof. Heat from the sun is absorbed by the water in the bags on the roof. This heat is in turn radiated downward through the metal roof, which also serves as the ceiling for the rooms below. The radiation is so even that no fans or blowers of any kind are required to distribute the heat throughout the home.

Automatic controls take care of all decisions. If rooms begin to overheat during the winter, the insulation panels automatically slide out to cover the waterbeds on the roof. If the outside weather is such that the bags are losing heat faster than they can collect it, again, the insulation slides over to cover the bags up. Seals prevent cold breezes from penetrating between the foam panels and the bags. During the night, of course, the panels close over the waterbags to prevent the escape of heat they have stored up.

The mass of water on the roof, plus the mass of the heavy concrete walls can store up about four days' worth of heat for comfort during a long cloudy spell. The house has no auxiliary heating system at all, and experience so far shows that none is needed.

Summertime cooling. In the summer, cooling of the home is accomplished by reversing the way things are done during the heating season. At night, the waterbags are exposed to the clear night sky. They lose heat to the sky by nocturnal radiation — the process of losing heat into the clear night sky. In Iran, people have been making ice in the deserts for centuries by means of nocturnal radiation. Water loses so much heat to the night sky that it will freeze, even when the temperature outside is as high as 48 degrees. You may have seen the same thing at home on a cool clear morning; you'll find frost on the ground even though the temperature is several degrees above freezing.

Once the water has been cooled it can absorb great quantities of heat from inside the home, directly through the metal roof. During the daytime, the insulation panels automatically roll out to cover the waterbags and to keep them from picking up any heat from the sun.

How well does all this work? Let's take a look at data based on inside and outside temperatures recorded year-round at Hay's first test house in Arizona. Even though winter temperatures outside dropped into the 20s, the temperature inside the home

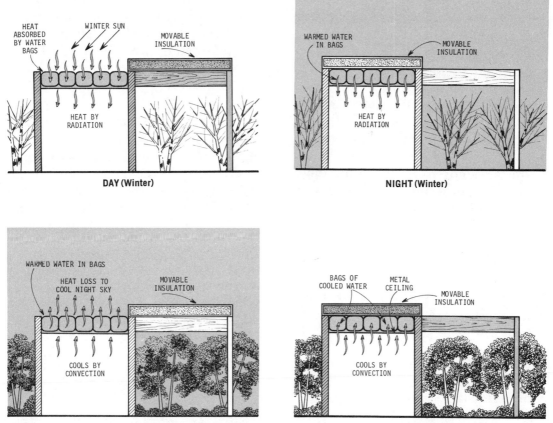

DAY (Winter)

NIGHT (Winter)

NIGHT (Summer)

DAY (Summer)

Harold Hay's Sky-Therm Process is the most elegant form of passive solar heating ever conceived. The system is based on rooftop waterbeds and movable insulation panels.

For winter heating, rooftop waterbeds are exposed to the sun, whereupon they are warmed. They release their heat down through the metal ceiling and heat the rooms below. At night, insulation is slid over the waterbeds to prevent heat loss.

For summer cooling, the opposite takes place. The beds are exposed to the night sky and radiate their heat into space. During the day the beds are covered with insulation to keep them cool while they soak up heat from the rooms below and thus cool the house.

During one year of operation at a Phoenix test site, this system maintained interior temperatures within a 69° to 80° F. range, while outdoor temperatures ranged from 20° to 113°. Monthly interior temperatures seldom varied more than 6°, while monthly outdoor temperatures routinely varied 50° to 60°. And the average interior temperature held in the mid-70s, except during coldest and hottest months.

As you can see, this is remarkable performance. Except for the electricity to move the motorized insulation panels, and to pump water onto the roof for evaporative cooling during the summer, no outside energy was required to keep the house comfortable. Passive climate control like this should be favored over more complicated systems whenever possible.

never went below 69 degrees. And even though summer temperatures often went over 100 degrees outside the home, inside, the temperature never went above 80 degrees.

All that from a home with no heating system, no fans, no airconditioner —nothing but some bags of water and movable insulation on the roof. Tenants who lived in another Hay-designed Sky-Therm house in California also got by without auxiliary heat or cooling and found the Sky-Therm cooling process more comfortable and, of course, quieter than conventional cooling techniques.

Price for a system like this? A group of scientists studying the system believe a home built with the Sky-Therm system would cost roughly the same as, or only slightly more than, a conventional home. Over the years, the economic and ecological advantages of heating and cooling for free would be staggering.

Drawbacks? The Sky-Therm process as described here is suited only to areas of the country where there are large diurnal temperature swings (large day/night temperature differences), and where snow build-up on the roof will not be a problem. But Hay has modifications of the Sky-Therm principle that can adapt it to any climate. In fact, that's the whole secret to Sky-Therm: It's designed to work with the climate, to put the climate to work rather than try to fight it with huge inputs of energy.

The Sky-Therm process is pat-

ented, but Hay will grant one-time rights to use it, and he'll offer consulting services to builders and architects who want to use the system. If you are interested, write Harold Hay at 2424 Wilshire Boulevard, Los Angeles, CA 90057.

Steve Baer's Drumwall System. Here's another passive solar heating system, similar in principle to Harold Hay's Sky-Therm system. It employs movable insulation and a large thermal mass of water, but in a slightly different way than Hay's system does. Baer and his family live near Al-

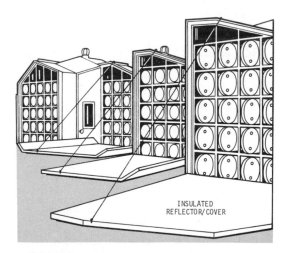

Steve Baer's drumwall-heated house features drums of water placed inside the all-glass south walls of four different rooms. The glass walls can be closed off by means of insulated, winch-up doors. Inner faces of these doors are foil faced to reflect extra light onto the drums when the doors are open and lying flat on the ground. This is a good example of passive solar design.

buquerque, New Mexico, in a single-story home consisting of a chain of domelike rooms called zomes. Not all of the rooms are heated by the sun, but those that are get about 90 percent of their heat from the sun. Note: This assumes you'll accept indoor temperatures down to about 55 degrees at certain times during the heating season.

The zomes are built on five-inch slabs of concrete that provide a great deal of mass for heat storage. Walls making up the exterior shell facing north, east, and west are sandwiches composed of aluminum skins with cardboard honeycomb and urethane foam between them. Interior walls, such as partitions, are adobe to provide even more mass. And many of the exterior sandwich walls are lined inside with adobe.

The south walls of the four rooms heated by the sun are completely glazed with a single layer of ordinary single-strength window glass. Inside the rooms, just behind these glass walls are big steel "wine racks" holding 55-gallon steel drums. Each drum is filled with about 53 gallons of water — to leave room for expansion — and painted black on the end facing the window. The other end of each barrel is painted white to help brighten the room. In all, a total of 90 drums hold about 20 tons of water. During the day, the sun shines in through the windows and heats the water in the drums. The water can store a great deal of heat, and release it slowly into the rooms, day and night. To further control the amount of heat passing from the drums to the room, curtains are used.

Outside is a large, winch-up insulating cover, lined on its inner surface with aluminum. This cover, hinged at its lower edge, can be closed over the glazed wall at night to prevent heat loss. During the day it is lowered to a position near the ground where its aluminum liner can reflect extra sunlight onto the blackened ends of the drums.

During the summer, things work in reverse. The cover is closed in the daytime to keep the sun off the drums. At night, cool air is circulated over the drums to cool them. Thus cooled, the drums can absorb heat from the room during the day. Baer reports that the inside temperatures of the zomes stay about 80 degrees even when the outdoor temperature rises over 100.

Supplementary heat comes from wood-burning stoves — a good means of solar backup. At present, the big insulating covers for the zomes are controlled manually by a hand crank and a nylon line. Of course the operation could be turned over to an electric motor and a system of temperature sensors.

While the Drumwall system is not quite as effective as the Sky-Therm system, it is more suited to northern climates. The glazed wall won't collect snow as will Hay's flat roof. Plans for the Drumwall are available from

Baer's Zomeworks. He also has plans for the Beadwall insulated window system described in Chapter 4. And he offers good plans for a domestic solar water heater—one you can easily make yourself. The address: Zomeworks, Box 712, Albuquerque, NM 87103.

Other Systems. Sky-Therm and Drumwall are the two best-known examples of simple, passive solar heating systems. And they are probably the most effective. But there are other

ideas in simple systems you might like to consider.

Steve Baer has also developed another solar heating system that consists of air collectors and a rock storage bin. Convection makes the whole thing work. Baer calls the design an air loop rock storage system. The system works like this:

A large area of air-cooled collectors is connected to a rock storage bin as shown in the accompanying drawing. The bin is located above the lower edge of the collector array. Air inside the collectors rises as it gathers heat. At the top of its rise the hot air enters the rock storage bin. As the air begins to lose its heat to the rocks, it descends through the rocks. When it reaches the bottom of the bin it is then ducted back to the bottom of the collectors to again pick up heat, rise, pass through the rock bin and dump its heat.

This is the same principle that drives a thermosiphon water heater. The force of the hot air rising through the collectors and the cool air dropping through the bin of rocks makes the whole thing go. These convective forces are not very great, however, so your design must allow for the freest flow of air possible. Baer advises that the collectors should have a great deal of surface area to transfer heat to the airstream. He says that several layers of expanded metal lath do a good job. (These are available from Kalwall.) The rock bin and ducts should be heavily insulated, with

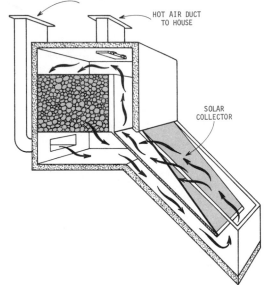

Here's the thermosiphon principle put to work in a solar hot rock storage loop. You can use a system like this if your house is built on a south slope. Build the rock bin under the house. Dampers on the ducts let you control heat coming into the house. Cover the collector during the summer to prevent unnecessary overheating. Steve Baer of Zomeworks recommends that all flow passages be at least $1/15$ as large as the collector area for free air flow. Insulate with six-inch batts, and make the collector at least six feet long. The longer, the better the air flow.

COLLECTOR UNDER WINDOW

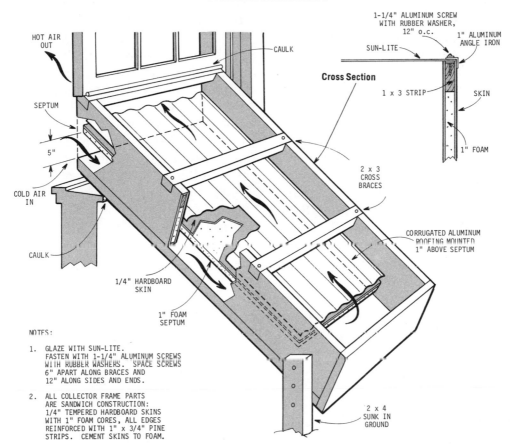

HOT AIR
OUT

CAULK

1-1/4" ALUMINUM SCREW
WITH RUBBER WASHER,
12" o.c.

1" ALUMINUM
ANGLE IRON

SUN-LITE

Cross Section

1 x 3 STRIP

SKIN

SEPTUM

1" FOAM

5"

2 x 3
CROSS
BRACES

COLD AIR
IN

CAULK

CORRUGATED ALUMINUM
ROOFING MOUNTED
1" ABOVE SEPTUM

1/4" HARDBOARD
SKIN

1" FOAM
SEPTUM

NOTES:

1. GLAZE WITH SUN-LITE.
    FASTEN WITH 1-1/4" ALUMINUM SCREWS
    WITH RUBBER WASHERS. SPACE SCREWS
    6" APART ALONG BRACES AND
    12" ALONG SIDES AND ENDS.

2. ALL COLLECTOR FRAME PARTS
    ARE SANDWICH CONSTRUCTION:
    1/4" TEMPERED HARDBOARD SKINS
    WITH 1" FOAM CORES, ALL EDGES
    REINFORCED WITH 1" x 3/4" PINE
    STRIPS. CEMENT SKINS TO FOAM.

2 x 4
SUNK IN
GROUND

The thermosiphon principle can work with air as well as it does with water. Here's an air-type collector put to use warming the air in a single room. You can fasten it into the window opening of any south-facing window. Since there is no back pressure on this unit (as there might be if you ran the hot air through a rock storage bin) it will thermosiphon quite effectively. Remember that the faster the air moves through the heater, the more efficient it will be. You can encourage air movement by insulating, as shown, and by keeping all passages as clear as possible. This rig will automatically stop circulating whenever it can't add heat to your room. No controls needed!

A simple unit like this can be quite effective since there are very small losses inherent in the design. Hot air exits the collector and is immediately where you want it, warming the house. Transmission losses are just about zero. So are storage losses. Figure a 4 x 8-foot collector will pump in somewhere around 25,000 Btu, during the course of a sunny winter day. This is about the same amount of heat you'd get from a quart of fuel oil, which means the unit provides only a little over a dime's worth of heat a day.

Construction is basically a foam and hardboard sandwich. The foam provides insulation and reinforces the skins. The skins in turn provide protection for the foam. Edges of all sandwich panels are reinforced with pine strips which also provide firm anchorage for the screws and nails that secure the frame parts and the glazing material. Cement the skins to the foam with a special foam cement. Note: ordinary cements may melt the foam.

Anchor the lower edge of the collector to the ground via 2 x 4 legs set into the ground. Make sure the anchorage is secure, or the wind could catch the collector and tear it loose.

Note: The absorber plate shown here is made of corrugated aluminum roofing. A slightly more efficient plate can be made of aluminum lath installed in four or five layers as illustrated early in this chapter.

large cross sections to avoid constricting the air flow.

A similar convective system without storage can be used for heating individual rooms in your home. Allow the collectors to vent into your home, and be sure to feed the collectors cool room air.

Another passive solar heating system is the one used by the French solar scientist F. Trombe and architect Jaques Michel. It is essentially the same type of idea that Steve Baer uses in his Drumwall. The Trombe/Michel design differs in two major ways, however. First, it makes no use of movable insulation, while Baer's design features a hinged insulation cover that can close off his large glass wall to prevent heat loss. Second,

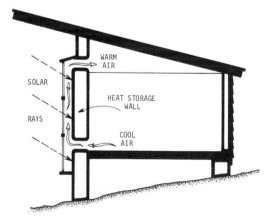

Trombe-Michel's solar wall is another simple passive solar heating system. A heavy concrete wall serving as absorber and heat storage sink sits just inside an all-glass south wall. Performance could be enhanced by addition of insulated doors like those described earlier for Steve Baer's Drumwall, or by the installation of Beadwall windows described in Chapter 9.

Trombe/Michel use a massive concrete wall instead of an array of water-filled drums. The wall is painted black on the surface that faces the sun, and it has air passages top and bottom to allow for a convection current to draw air through the heating zone, as shown in the accompanying drawing. The bottom duct openings are located above the bottom of the glass collecting area, so that a reverse convection current won't be established during the night and periods of cloudiness.

In summer, vents to the outdoors, located at the top of the solar wall, allow heated air to pass out of the home. This creates a thermal chimney effect, which can then draw air through the house and provide a certain amount of cooling.

According to Trombe, a system of this kind will heat ten cubic feet of living space for every square foot of solar wall. Two test houses built in the Pyrenees got about 60 percent of their heating needs from the Trombe/Michel solar wall. This performance could probably be improved by adding a movable insulated cover such as Steve Baer uses on the Drumwall, or by making the entire southern wall of the home out of Beadwall instead of plain glass.

Substituting a masonry wall for Baer's drums has advantages. The masonry will certainly look more attractive to most eyes than will a stack of steel drums. And a masonry wall is unlikely to spring a leak. On the other side of the coin: A masonry wall is not

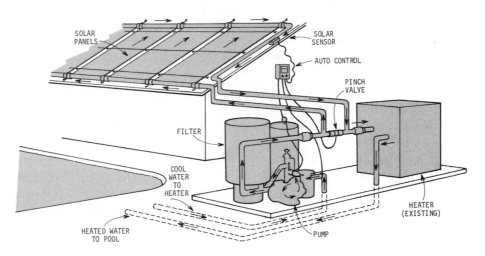

Here is a typical Fafco swimming pool heater installation setup. This system includes an auto-control center to handle switching from solar to conventional heating modes. And the existing filter pump circulates water through the plastic collector panels.

as efficient a heat sink as a Drumwall; and thus thermal performance is lower.

SOLAR SWIMMING POOL HEATERS

For the sun, it's an easy task to keep the water in your pool warm. Often, the solar pool heater only needs to raise the temperature of the pool water a few degrees to make it comfortable. This eliminates the need for elaborate solar collectors. The basic commercial solar collector for swimming pools is made of black plastic, has no glazing, no insulation, and costs only about two bucks a square foot.

Fafco, one of the first makers of plastic pool heaters, recommends a collector area equal to approximately 50 to 75 percent of the area of your pool. This means a small pool 15 by 30 feet would need something like 250 square feet of collector area. This would cost around $500 or $600. If your pool is in a particularly windy area, or is shaded, you might need to increase collector area to equal the surface area of your pool. But few pools are located in the shade, and most pools are fenced, which tends to temper winds.

Aside from collectors, there's not much else required for solar pool heating. The pump that runs your pool filter can also feed your collectors. Run some plastic pipe from your present filter lines up to the collectors and back to the pool, and the installation is nearly complete.

Controls? The simplest way to go is merely to set your filter pump to run only in the daytime. This eliminates the need for any other controls, and will give good results during the heart of the swimming season. Early in the season, and again at the tail of the season, air temperatures are low enough to cause trouble with a simple system like this. If the weather is cloudy, water pumped through the collectors will be cooled by the surrounding air; the system will actually lose heat.

So if you want your heater to provide the longest swimming season possible, use a differential thermostat to control the pump. One probe should go at the collectors, the other should sense the temperature of the pool water. As long as the collectors are warmer than the pool water, the pump will run. A simple thermostat to handle this task will cost from $50 to $100, and it's a good investment once you've put several hundred into collectors.

There is one other way to control the pump: Use a sun-sensing electric eye to switch the pump on. To work effectively, the photoelectric cell must be calibrated to do its switching only when the sky is clear or slightly overcast. It shouldn't trigger the pump on a heavily clouded day.

As for simple pool heating tricks, there are a couple of very simple ways you can help the sun heat your pool without installing collectors. You could paint the inside of the pool a darker shade than the customary light blue. Black would be the best color for solar absorption, of course, but it may not appeal to your eye. In that case, consider a very dark blue. This color was used for the diving pool at the Montreal Olympics. It looks good, giving the pool a very deep look.

Another simple pool heating solution is the Catel Solar Circle. It's nothing more than a five-foot circle of polypropylene plastic that floats on the surface of your pool. To some extent, the plastic helps retain heat in the pool, much as the clear cover on a solar collector. Most of its benefit comes from the fact that it stops evaporation of water from the area it covers. This puts an end to evaporative cooling and helps keep the pool warm. Fred Rice Productions, distributor of Solar Circles, claims they can raise the temperature of a pool about one degree a day up to a limit around 80 degrees if used in sufficient numbers. Rice recommends covering about 75 percent of your pool area with the circles. Each circle has an area of almost 20 square feet, so 15 of them would be required for a 15x30-foot pool. At $10 a circle that would come to $150.

You can remove and stack the circles whenever you want to swim, and leave them in the pool at other times.

Left in the pool, the circles can serve as pool toys for children. Note also that the circles have beneficial side effects: Since they cut water loss due to evaporation, they thereby cut chemical costs too.

SOLAR AIRCONDITIONING

There's not much you can do to cool your house with solar energy. While it is possible to use the sun's heat to drive an absorption-type chiller, all such chillers at present are sized for commercial applications; they're just too big for cooling a home.

There are, however, a few tricks for cutting cooling costs. They aren't really solar cooling, but they work. For example, Solaron's air-based solar heating sytem can provide summer cooling by drawing nighttime air into the rock storage bin. This cools the rocks. During the daytime, house air can be blown through the rock bin to be cooled.

Harry Thomason uses a different approach. He cools his rock bin with an airconditioner. But he runs it at night when utility loads are low, and when air is cool and the airconditioner can run most efficiently. Then during the daytime he blows room air through the chilled rock bin for cooling. This system significantly increases the cooling power of the airconditioner. It can offer further savings in areas of the country where electricity is available at a reduced rate during off-peak hours. Lower rates for off-peak use will probably

become more prevalent as time goes by, but right now the practice is mostly experimental.

ELECTRICAL ENERGY FROM THE SUN

What are the possibilities of using solar cells to generate your own electricity at home? You could do it, but the cost at present is much too high. Right now, the cost of electricity from solar is about 30 times higher than it will have to be to become practical. Today's solar cells are now down to about $15 per watt. And the price continues to get lower. But the price won't be reasonable until it gets down to around 50 cents a watt. And that probably won't happen until the late 1980s.

SOURCES FOR SOLAR EQUIPMENT
Air-cooled Solar-collector Makers

International Solarthermics
Box 397
Nederland, CO 80466

Solaron Corp.
4850 Olive St.
Denver, CO 80022

Sunworks
Box 1004
New Haven, CT 06508

Water-cooled Solar-collector Makers

CSI Solar Systems Div.
124000 49th St. North
St. Petersburg, FL 33732

Corning Glass Works
Corning, NY 14830

E & K Service Co.
16814 74th Avenue N.E.
Bothell, WA 92075

Ecotechnology
234 Barbara Ave.
Solana Beach, CA 92075

Energex Corp.
481 Tropicana Rd.
Las Vegas, NV 89109

Energy Conservation Institute
3130 Spring St.
Redwood City, CA 94063

Energy Systems, Inc.
941-D Anistad Court
El Cajon, CA 92020

Fafco (swimming pool heaters)
138 Jefferson Dr.
Menlo Park, CA 94025

Free Heat
Box 8934
Boston, MA 02114

Fun and Frolic (swimming pool
    heaters)
Box 277
Madison Heights, MI 48071

Garden Way Labs
Box 66
Charlotte, VT 05445

Grumman Aerospace Corp.
South Oyster Bay Rd.
Bethpage, NY 11714

Martin-Marietta Corp.
11300 Rockville Pike
Rockville, MD 20852

PPG Industries
One Gateway Center
Pittsburgh, PA 15222

Piper Hydro, Inc.
2895 E. La Palma
Anaheim, CA 92806

Raypak, Inc.
31111 Agoura Rd.
Westlake Village, CA 91361

Revere Copper and Brass
Box 151
Rome, NY 13440

Reynolds Metals
2315 Dominquez St.
Torrance, CA 90508

Solar Power Corp.
930 Clocktower Pkwy.
New Port Richey, FL 33552

Solarsystems, Inc.
1802 Dennis Drive
Tyler, TX 75701

Sun Systems, Inc.
Box 155
Eureka, IL 61530

Sundu Co. (swimming pool heaters)
3319 Keys Lane
Anaheim, CA 92804

Sunsource
9606 Santa Monica Blvd.
Beverly Hills, CA 90210

Sunworks
Box 1004
New Haven, CT 06508

Tranter, Inc.
735 E. Hael St.
Lansing, MI 48909

Solar Water-heater Makers

American Heliothermal
3515 Tamarac
Denver, CO 80237

Aztec Solar Co.
2031 Dyan Way
Maitland, FL 32751

Beasley Solapak
Solar Energy Research
Box 17776
San Diego, CA 92117

C.B.M. Mfg.
621 N.W. 6th Ave.
Ft. Lauderdale, FL 33311

Capital Solar Heating
475 N.W. 25th St.
Miami, FL 33127

Energy Systems Inc.
634 Crest Drive
El Cajon, CA 92021

Hitachi America Ltd.
437 Madison Ave.
New York, NY 10022

SAV
Fred Rice Productions
6313 Peach Ave.
Van Nuys, CA 91401

Sol-Therm Corp.
7 W. 14th St.
New York, NY 10011

Sola-Ray
Largo Solar Systems
2525 Key Largo Lane
Ft. Lauderdale, FL 33312

Solector
Sunworks
Box 1004
New Haven, CT 06508

Sunsource Pacific
501A Cooke St.
Honolulu, HI 96813

Sunstream
Grumman Energy Programs
Box 365
Bethpage, NY 11714

Wilcox Mfg.
13375 US 19 N.
Box 455
Pinellas Park, FL 33565

**Catalog Sources of Solar Components
(Controls, pumps, fans, etc.)**

A-Z Solar Products
200 E. 26th St.
Minneapolis, MN 55404

Garden Way Labs.
Box 66
Charlotte, VT 05445

General Energy Devices
2991 West Bay Drive
Largo, FL 33540

Kalwall Corp.
Solar Components Div.
Box 237
Manchester, NH 03105

Solar Research Div. of Refrigeration
  Research
525 N. 5th St.
Brighton, MI 48116

WIND POWER

The first thing you should know about electric power from the wind is that it doesn't come cheap. If you think you can set up your own wind-generating plant that will compete with the local power company, forget it. An installation capable of providing the kind of service you get from the power company is going to cost you at least $15,000. As soon as this simple economic fact sinks in, most wind-power dreamers give up their dream. And over the past few years, a lot of dreams have been dashed.

Take the case of Solar Wind—a U.S. pioneer company in supplying wind generating equipment, and now part of Enertech Corporation. During the first 2½ years Solar Wind was selling wind equipment, they received almost 30,000 requests for information on their products. Out of those 30,000 requests came just over 50 sales. From this it's reasonable to conclude that of all the people interested in wind generation, only a small percentage actually invest in it.

WHEN WINDPOWER MAKES SENSE

We've painted a dark picture so far. But there is a bright side, too. Although most of those 30,000 people who contacted Solar Wind decided not to turn to wind power, over 50 of them did. There are good reasons for installing a wind plant. Henry Clews of Enertech's Solar Wind says:

"If your location is more than 2,000 feet from the nearest power pole, you may be able to justify a wind-power system from a purely economic standpoint. Most of the systems we have sold and installed to date fall into this category. These include remote sites, islands, country homes, camps and farms, lookout stations and communications repeater stations, all having good access to wind, and poor access to power lines. A wind electric system can look very appealing when compared to a noisy, costly and troublesome gasoline generator.

"So, if you are in a situation with good access to wind, and poor access to power, wind electrical plants may be right for you. They are a definite improvement over gasoline or gas-powered generators."

One of the first wind systems Solar Wind sold went to Neil Welliver, a man in Maine who wanted to replace his 5,000 watt L.P. gas generator. The gas generator was costing him over $50 a month in fuel. And it was a source of constant service headaches. Most reasonably-priced, engine-driven generators require an overhaul after little more than a month of use. A good wind system can run for years with virtually no attention.

A wind system is much quieter than an engine-driven generator. And if you've gone to the bother of living so far from civilization that you have no easy access to commercial power, quiet probably means a lot to you. Chances are, you are also interested in the environment and in sound en-

ergy use. A wind plant will burn no fuel and cause no pollution. And compared to a fuel-burning generating system of your own, it can even save you money. Neil Welliver figures his wind plant will pay for itself in three or four years, compared to his gas generator costs.

THE BASIC WIND SYSTEM

It takes more than a generator hooked to a propeller to produce a complete generating system. Starting from the

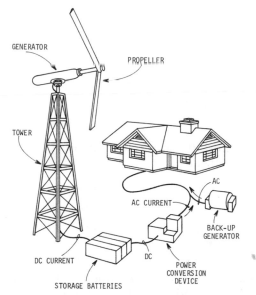

Basic elements of a windpower installation include a propeller and generator mounted on a tall steel tower. DC current from the generator is stored in lead-acid storage batteries designed for the purpose. These are not ordinary car batteries, but rather special types that can stand repeated deep discharging. Power from the batteries is converted to AC for most home appliances by means of an inverter. A backup generator is required to supply power in periods of calm when storage batteries become drained.

PREVAILING DIRECTIONS AND MEAN SPEED (mph) OF WIND: ANNUAL

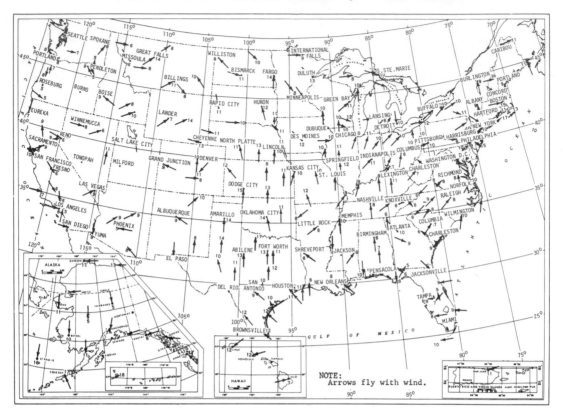

The map above is one of many on the wind contained in the *Climatic Atlas of the United States*. The map shown here indicates average annual speeds and prevailing directions.

top and working down you'll need the following:

1. Generator and propeller.

2. A tower to place the generator up in the air where it can receive the full power of the wind.

3. A storage system of lead-acid batteries. The wind doesn't blow all the time. When it doesn't, you live off the power stored in the batteries.

4. Power converters. The current stored in your batteries is DC. But most appliances are designed for AC. You need some type of converter to change that DC current to AC.

5. A backup system to provide electricity when your storage system runs dry during a long calm spell.

As it turns out, much of the cost of a wind system goes into these additional elements. Together with the labor required to install a wind system, they add up to more than the base cost of the actual propeller and generating equipment. But if you are willing to do your own work you can cut costs considerably. And if you can make other compromises, such as settling for DC and its limitations and getting by without a backup system,

BEAUFORT NUMBER	WINDSPEED M.P.H.	MEAN WIND-FORCE LB/FT²	QUALITATIVE DESCRIPTION OF WIND
0	0-1	0	Calm. Smoke rises vertically.
1	1-3	0.01	Light air. Smoke drifts but wind vanes do not turn.
2	4-7	0.08	Light breeze. Wind vane turns. Leaves rustle.
3	8-12	0.28	Leaves and twigs in constant motion.
4	13-18	0.67	Moderate breeze. Raises dust and loose paper.
5	19-24	1.31	Small trees in leaf begin to sway. Small crested waves form.
6	25-31	2.3	Strong breeze. Large branches move. Telephone wires whistle.
.7	32-38	3.6	Moderate gale. Large trees in motion. Walking is difficult.
8	39-46	5.4	Strong Gale. Extreme difficulty in walking against wind.
9	47-54	7.7	Light roofs liable to blow off houses.
10	55-63	10.5	Hurricane. Even the strongest mills liable to be damaged.

Fairly accurate estimates of windspeed without the use of instruments are possible with the information in the table above.

you can cut costs even more. But more about that later. First, let's take a good look at a complete system.

The Propeller and Generator. These two elements are the heart of your system. They capture the energy of the wind and convert it into useful electricity. The amount of juice they produce is determined by four simple variables.

First, there's the speed of the wind. Wind speed is very important, and the faster (up to a point) the better. This is because the power available from the wind is proportional to the cube of the wind speed. Double the speed and you get *eight* times as much power. Too much wind, however, can overstress a windplant.

The second variable is the size of the propeller. The power a prop can extract from the wind is proportional to the square of the prop's diameter. If you double the diameter you get four times as much power.

The third variable is the size of the generator. Naturally, the generator should be matched to the winds and the prop. A generator that's too small for your local wind conditions, or too small for your propeller, cannot produce up to the full potential of your site. On the other hand, a generator that's too large for your prop and wind conditions is a waste of money, and it will cost you power at low wind speeds.

The fourth variable is the efficiency of your propeller. Wind generators use two or three-bladed props that look a lot like airplane props. As it turns out, most modern windplants are roughly equal in efficiency. There are no secret propeller formulas that give one manufacturer a big edge over any of the others.

HOW TO PICK A WINDPLANT

This is basically a matter of choosing a windplant that will supply you with enough power to suit your needs. Because of the variability of the wind, and of your electrical consumption, truly accurate calculation of the size mill you need is difficult. On the other hand, truly accurate sizing isn't very important. What you need is enough information to be sure you get a system that is sized fairly close to your needs. And this isn't too hard to do.

Start by assessing your wind situation. Write to the Environmental Data Service, National Climatic Center, Federal Building, Asheville, NC 28801. Ask for their *Climatic Atlas of the U.S.* (Price is about $6). This is a spiral-bound, 80-page collection of maps and charts with a special wind section that shows strongest winds, monthly average (mean) speeds and prevailing direction for hundreds of weather stations across the U.S. Check the station closest to you, paying special attention to the average yearly speeds. It's a sort of rule of thumb that winds in your area should average at least 10 mph if you plan on getting a useful amount of power from a windmill.

If the readings for your nearest station don't reach 10 mph, don't give up. Wind speeds can vary widely even over very short distances. Your next step is to check wind speeds right at your site. To do this, you'll need a wind meter. For $7.50 you can get a nice little hand-held meter by Dwyer Instruments that will do the job. A more sophisticated version by Dwyer will cost you $30.95. Both these prices are postpaid from Solar Wind Division of Enertech, P.O. Box 420, Norwich, VT 05055.

For the most meaningful results, you should make your readings exactly where you plan to place your mill. This will be in an area free from surrounding obstructions, and as high in the air as possible. This is where the more expensive meter comes in handy. Its sensor is fastened at the end of 50 feet of flexible tubing. This means you can secure it to the top of a temporary mast to get it up in the air where your mill will go, yet you can read the gauge from the ground. If you use the handheld meter, you'll have to climb a tree or whatever to get it up to the right height. And, you'll have to do this at least once or twice a day for several weeks—for a year if possible.

The object is to get a series of daily readings taken at roughly the same times every day. While you're at it you might take the readings at several prospective mill locations to find out which one is best. Remember that the power in a wind increases as the cube of the windspeed, so even a slight advantage in speed can mean a significant increase in the power.

I mentioned earlier that you should make your tests for at least a few weeks, or a year, if possible. The longer you run your tests, the better the chances that they'll represent the true picture of wind speed in your area. If you want to rush your testing you run the risk of bad results. For example if you take all your readings during a two-week period of abnormally high or low winds, your results won't have much meaning.

That's where the climatic atlas I mentioned earlier comes in. Now you can compare the results you get from your tests with the figures given for a nearby weather station listed in the atlas. If your figures seem to agree fairly well with those in the atlas, you know you must be doing something right. But if your figures are all out of line with those in the atlas, you should do some more testing of your own. Once you've satisfied yourself that you have a good idea of your local wind conditions, half the battle is over.

Now you have to determine your electrical needs. The simplest way is to base this on the electrical consumption noted on your bills over the past year. The accompanying table showing electrical consumption of household appliances will help estimate consumption of specific appliances, should you plan to buy additionals or build a new house. It lists average power consumption in watts, the length of time they are used per month, and their total consumption in KWH per month. These figures are based on averages. If you want, you can get more precise figures by substituting the watt ratings of your own appliances (the ratings are usually indicated on a specifications label) and the average monthly usage for those appliances in your own home. To figure your own usage you can log the time you run some of the appliances and then estimate others. If you follow this procedure, multiply watts on the label times hours used per month. This gives watt hours per month. To get kilowatt hours per month just divide by 1,000. Do this by simply moving the decimal point three places to the left.

ELECTRICAL CONSUMPTION OF HOUSEHOLD APPLIANCES

APPLIANCES	POWER WATTS	CURRENT REQUIRED AMPS at 12v	at 115v	TIME USED HRS/ MO.	TOTAL KWH PER MO.
Airconditioner (window)	1,566	130	13.7	74	116.
Blanket, electric	177	14.5	1.5	73	13.
Blender	350	29.2	3.0	1.5	0.5
Broiler	1,436	120.	12.5	6	8.5
Clothes Dryer (electric)	4,856		42.0	18	86.
Clothes Dryer (gas)	325	27.	2.8	18	6.0
Coffee Pot	894	75.	7.8	10	9.
Dishwasher	1,200	100.	10.4	25	30.
Drill (¼ in. elec.)	250	20.8	2.2	2	.5
Fan (attic)	370	30.8	3.2	65	24.
Freezer (15 cu. ft.)	340	28.4	3.0	290	100.
Freezer (15 cu. ft.) frostless	440	36.6	3.8	330	145
Frying Pan	1,196	99.6	10.4	12	15.
Garbage Disposal	445	36.	3.9	6	3.
Heat, electric baseboard, average size home	10,000		87.	160	1600.
Iron	1,088	90.5	9.5	11	12.
Light Bulb, 75-watt	75	6.25	.65	120	9.
Light Bulb, 40-watt	40	3.3	.35	120	4.8
Light Bulb, 25-watt	25	2.1	.22	120	9.
Oil Burner, ⅙ hp	250	20.8	2.2	64	16.
Range	12,200		106.0	8	98.
Record Player (tube)	150	12.5	1.3	50	7.5
Record Player (solid st.)	60	5.0	.52	50	3.
Refrigerator-Freezer (14 cu. ft.)	326	27.2	2.8	290	95.
Refrigerator-Freezer (14 cu. ft.) frostless	615	51.3	5.35	250	152.
Skill Saw	1,000	83.5	8.7	6	6.
Sun Lamp	279	23.2	2.4	5.4	1.5
Television (b & w)	237	19.8	2.1	110	25.
Television (color)	332	27.6	2.9	125	42.
Toaster	1,146	95.5	10.0	2.6	3.
Typewriter	30	2.5	.26	15	.45
Vacuum Cleaner	600	50.5	5.5	6.4	4.
Washing Machine (auto)	512	42.5	4.5	17.6	9.
Washing Machine (wringer)	275	23.	2.4	15	4.
Water Heater	4,474		39.	89	400.
Water Pump	460	38.3	4.0	44.	20.

WINDMILL SIZING CHART

	MONTHLY OUTPUT IN KILOWATT HOURS					
NOMINAL OUTPUT RATING OF GENERATOR IN WATTS	AVERAGE MONTHLY WIND SPEED IN MPH					
	6	8	10	12	14	16
50	1.5	3	5	7	9	10
100	3	5	8	11	13	15
250	6	12	18	24	29	32
500	12	24	35	46	55	62
1,000	22	45	65	86	104	120
2,000	40	80	120	160	200	235
4,000	75	150	230	310	390	460
6,000	115	230	350	470	590	710
8,000	150	300	450	600	750	900
10,000	185	370	550	730	910	1,090
12,000	215	430	650	870	1,090	1,310

The two tables above give you the information you need to pick the right size windplant, and to size your battery storage system. See the accompanying text for methods. Tables are reprinted with permission from Henry Clews' excellent booklet "Electric Power from the Wind," available at $2 from the Solar Wind Division of Enertech.

AVAILABLE WINDPLANTS

MAKER	MODEL	OUTPUT WATTS	OUTPUT VOLTS	ROTOR DIAMETER IN FEET	COST	APP. DOLLARS PER WATT	NOTES
AEROPOWER	A	1,000	12	6	$1200.	$1.20	Larger models available from Prairie Sun and Wind.
BUCKNELL		220	12	5	$900. w/tower	$4.00	Available in kit form.
DOMINION	15	6,000*	N/A	15	$4600.	$.75	Darrieus type, vertical-axis designs.
	20	12,000*	N/A	20	$9000.	$.75	
DYNA	Wincharger	200	12	6	$390.	$2.00	Includes 10 ft. tower. Other voltages available.
ELECTRO	WV15G	1,200	12/115	9.8	$3500.	$3.00	WVG50G is most practical big windplant available. It produces the most power at the lowest cost.
	WV253G	2,500	115	12.5	$4250.	$1.75	
	WVG50G	6,000	115	16.5	$6580.	$1.00	
LUBING		400	24	7.2	$2950.	$7.40	Also make pumping mills.
SENCEN-BAUGH	500-14	500	12	NA	$2490.	$5.00	Very reliable.
	1000-14	1,000	12	NA	$2990.	$3.00	
DUNLITE	M	2,000	24/115	12	$3475.	$1.75	Most popular windplant sold in U.S. (Very reliable)
KEDCO	1,200	1,200	14.5	12	$1695.	$1.50	Plans available from Helion.

* Estimated output

Note: Aerowatt S.A. of France also makes windplants, but at $4 to $120 per watt they are priced out of the range of homeowners.

Okay, add up all the kilowatt-hours-per-month figures for all your appliances. This will tell you how much electricity you'll need per month. Now you can find out what size mill you'll need. Look in the windmill sizing chart on the preceding page under the column that corresponds to your average local windspeed. Find the KWH rating that comes out slightly above the KWH-per-month total you've just computed. Then follow that row over to the right to see what the watt rating of your generator should be.

Example: You total up all your electrical needs for a month and find it comes to 500 KWH per month—slightly less than the national average. You've already checked your average windspeeds and found them to be 10 mph. Again, go to the windmill sizing chart and read down the 10 mph column. You'll see that a mill rated at 10,000 watts will give you 550 KWH per month. You only need 500, but those extra 50 will help take care of the inefficiencies built into your system (mainly losses in the storage system).

Now you know what you need—a mill rated at roughly 10,000 watts. If you take a look at the table of available wind plants you'll have a rude awakening. There are no commercially available windplants rated as

SELLERS OF
WINDPOWER EQUIPMENT

SELLERS	MAKES
Budgen and Associates 72 Broadview Ave. Pointe Claire, Quebec, Canada	Lubing Elektro
Bucknell Engineering 10717 E. Rush St. South El Monte, Cal. 91733	Bucknell
Dominion Aluminum Fabricators, Ltd. 3570 Hawkestone Rd. Mississauga, Ontario Canada L5C 2V8	Dominion Darrieus-type windplants
Enertech Corp. Box 420 Norwich, Vt. 05055	Dunlite Elektro
Environmental Energies 11350 Shaefer St. Detroit, Mich. 48227	Dunlite Elektro
Helion PO Box 4301 Sylmar, Cal. 91342	Kedco
Independent Power Developers Box 618 Noxon, Mt. 59853	Dunlite Elektro Wincharger
Prairie Sun and Wind 4408 62nd St. Lubbock, Tex. 79414	Aero Power Wincharger Others
Sencenbaugh Electric Box 11174 Palo Alto, Cal. 94306	Dunlite Elektro Sencenbaugh
Aeromotor Water Systems Broken Arrow, Ok. 74012	Aeromotor water pumping mills
Dempster Industries Box 848 Beatrice, Neb. 68301	Dempster water pumping mills
Windworks Box 329, Rte 3 Mukwanago, Wis. 53149	Gemini inverter plans

Note: Most suppliers also stock accessory items such as towers, batteries, inverters, and weather instruments.

high as 10,000 watts. You now have two choices. You can install two mills such as the big 6,000 watt Elektro, or you can resolve to rethink your future electrical consumption with an eye towards cutting it roughly in half. Then you'll be able to get by with a single mill.

CHOOSING STORAGE EQUIPMENT

So far, you've only sized your mill. Now you have to decide how much storage you'll need. Go back to the chart labeled "Electrical Consumption of Household Appliances." You'll notice that in addition to the information on watts, operating times, and monthly consumption, all items show current required in amps. These columns list the amps required for both 12 and 115 volt systems. So you can figure what you need depending on the voltage you decide to use.

Two factors determine your storage needs. Most important is the total

The big Elektro WVG 50G, shown here on Henry Clews' Maine homestead, is the largest windplant you can buy today. It produces a maximum of 6,000 watts. It is also one of the most practical, and produces more for the money than any other mill. Even so you'd probably need a pair to supply you with the same amount of electricity used by the average American family. A total system of two mills might cost you as much as $20,000.

amount of energy you'll want to store away for use during calm periods. The more storage capacity you have, the longer you can hold out without wind, and the less you'll have to depend on backup power systems. Three day's worth of storage is a workable minimum. If you can increase this to about a week's worth you may never even need a backup system. To determine how much storage you need, divide your total monthly KWH consumption by 30 to find out how much power you use per day. Now convert that figure to amp-hours using the formula ah = wh/volts. (Amp hours equal watt hours divided by the voltage you're choosing.) The result will be the amp-hours you use per day. A storage system capable of storing four days' worth of power would then have four times the amp hours of the figure you've just computed. A week-long storage system would have seven times that.

Example: Let's go back to our earlier calculation of windmill size. In that example we found our monthly KWH consumption to be 500. Divide that figure by 30 to get daily consumption. Result: 16.6 KWH, or 16,600 watt hours. Now using the formula ah = wh/volts we get ah = 16,600/115 (assuming we're going to use a 115-volt system). The result is 144 amp hours per day. Four times that figure would equal 576 amp hours (four days' storage capacity).

Now for the second factor in sizing your storage. All batteries have a limit to the amount of current they can release at a given time. This is called the maximum discharge rate. Let's say you check your battery specifications and find that they have a 20-amp maximum discharge rate. This means you should never try to draw more than 20 amps from them at a given time. So go back to the appliance consumption chart and check to see which of your appliances might conceivably be on line at the same time. Remember that some appliances will turn on independently of your instructions, while others are under your control. For example, oil burners, refrigerators, freezers, and well pumps all switch on and off according to orders from thermostats or pressure-sensitive switches. You never know when they all might come on at the same time. Other appliances are under your control, and you can see to it that your family uses no more than one or two at a time in order to keep the battery discharge rate below its maximum.

So, let's say you have a storage system with a maximum discharge rate of 15. Your home has a refrigerator that draws 5.35 amps, and an oil burner that draws 2.2 amps. You have no control over these two appliances (unless you want to turn them off), so you have to assume that they may operate at the same time. The total of these two would be 7.55 amps. If you decided to use an electric frying pan rated at 10.4, and while you were cooking, both the refrigerator and the oil burner switched on, your total discharge rate would be 17.95. This would exceed your 15 amp maximum discharge rate, and you'd blow a fuse. This shows the importance of knowing the maximum discharge rate of your batteries, and the number of amps your appliances draw.

In the example above, we have a problem: We can't cook, heat our home, and run a refrigerator all at the same time. Solutions? You have two of them. You can turn off the refrigerator while you use the frying pan. Thus there's no danger that it can cut in while you're cooking (and while the oil burner is running). And there's no danger of blowing a fuse. When you finish cooking, just remember to turn the refrigerator back on.

This is a minor inconvenience, but it may appeal to you more than buying a second set of batteries to boost your maximum discharge rate to the point where it can handle larger loads. Buying extra batteries is your second solution to the problem of overloading.

POWER CONVERTERS: AC, DC, OR BOTH?

Not all electricity is created equal. Most appliances are designed to run on the current supplied by power companies. This is 60-cycle AC at 115 volts. But any wind generating system you can buy is going to supply not AC, but DC. Sure most modern wind systems use alternators which produce AC. But this AC is run through a rectifier and converted to DC so that it can be stored in batteries. So even if you buy a system with an alternator (instead of a DC generator) you'll still end up with DC.

This means you'll need a device to convert that DC from the batteries to AC that's palatable for your appliances. One of the cheapest ways to do this is with a rotary inverter. This is simply a DC electric motor that runs at a constant speed off battery power. It in turn drives an alternator which produces 115-volt AC — the stuff you and your appliances want. Rotary inverters produce current just like power company current, complete with a smooth sine wave. This is important if you want to run electronic devices on wind power. Appliances such as TVs and stereos all demand a smooth sine wave for static-free performance.

Another way to convert DC to AC is with an electronic inverter. These cost more than rotary inverters, and they usually don't produce a smooth wave — although some do. That's two strikes against them. Their advantage is efficiency. An electronic inverter may be around 80 to 98 percent efficient in its job of turning DC into AC, while a rotary unit may be only 60 percent efficient.

What does all this mean? It means you have to make some decisions about how you want to handle your electricity once you have it produced and stored. Here's the best plan of attack I've seen. It's the one used by Henry Clews at his wind-powered home in Maine.

First of all, you should arrange your storage batteries so they'll supply you with 115-volt DC. This means you'll want to use nine 12-volt batteries. (9 times 12 volts is 108 volts, close enough to 115 to do the job.) This 108-volt DC will do a lot of work just as it is—unconverted. For example it will run all your incandescent lighting fixtures. It will run all appliances with simple heating elements, such as percolators and frying pans. It will run small power tools with universal (AC/DC) motors. And it will do all this without conversion, or conversion losses.

Of course this doesn't do your AC appliances any good. They still need that AC power. So you use an inverter to give it to them. But in order to supply an AC diet for some appliances, and DC for others, you'll need a dual wiring system. Clews uses a dual system, with dual sets of wires, and dual outlets, labeled AC and DC. The DC outlets are always live. But the AC outlets only provide current when the rotary inverter is switched on by hand. Clews chose this procedure because the inverter uses power even when no AC from it is being consumed. There's no point in running the inverter when it isn't producing any AC.

Now Clews even goes a step beyond this. He uses two inverters. One is a 250-watt Navy surplus item, the other is a smaller 75-watt inverter. He uses the large one to run large appliances such as his TV, the smaller one to handle his radio and electric typewriter. This gives him greater efficiency than he'd get by running everything off the one large inverter. For example, he found that his small appliances were effectively consuming three times their normal wattage when fed by the large inverter, just because so much power was being wasted driving the oversize inverter.

If you want a simpler way to solve this inefficiency problem, you can use a solid-state inverter. As mentioned earlier, these are more efficient than rotary inverters. A big 3,000-watter draws only about 60 watts at no load, while Clews' little 250-watt rotary inverter uses 100 watts at no load. Just remember you can expect to pay something like $2,000 for a solid-state inverter of this size.

What about straight 12-volt DC power? Thanks to the motorcamping industry, 12-volt DC can be fairly useful. There is a good variety of appliances for the RV trade that will run on 12 volts. Some of these include incandescent and fluorescent lighting fixtures, refrigerators, pumps, air compressors, fans, coffee pots, and TVs. Some makers (Astral for one) make refrigerators that run on LP gas, and on 12 or 115 volt current. Of course these units are fairly small, and they cost quite a bit more than conventional refrigerators (about $550 for a seven cubic footer) but the gas/electric feature is one way around a backup generating system.

Don't forget that any car radio, tape player, or CB will also run on 12-volt

DC. And there's no reason why you can't use them in your home. Any good car stereo system will give you good performance at home, as long as you don't expect to achieve high volume levels; car stereos are designed to fill a relatively small space, such as a car interior, with sound and they may not have the power to fill a large home.

EXCHANGING ELECTRICITY WITH YOUR POWER COMPANY

There's one other way to handle the whole problem of providing your appliances with the right diet. And at the same time it solves the problem of energy storage, by eliminating the need for storage batteries. This new alternative is the Gemini Power Conversion Unit, made in Wisconsin. Here's what it does:

First, it converts DC to 60 cycle AC at 120 or 240 volts, just as an ordinary solid-state inverter can do.

Second, it connects you directly to power company distribution lines. If your mill is generating no power, or not enough power to meet your needs, the Gemini converter draws power from the utility to make up the difference.

If your mill is generating just as much power as you need, the Gemini cuts off utility power and you run strictly on your own juice.

If you happen to be producing more power than you can use, Gemini shunts the excess back into the power company lines, and your electric meter runs backwards. The ex-

cess power enters the power pool of the utility company system where it is used by any of the customers hooked up to that line.

Here are the advantages of the system. Even though the Gemini is expensive (about $1,200), it eliminates the need for a battery storage system, and for a backup system. The power company serves both these needs. Excess power is, in effect, stored in the power company grid. Whenever your wind system can't supply you with enough power the utility can, acting as a backup system. Eliminating those two systems saves a lot of money. According to Windworks, makers of the Gemini, "Interfacing the Gemini with off-the-shelf wind generating equipment brings the cost of wind power down to approximately 10 cents per kilowatt hour when the system cost is amortized over the life of the equipment." So, despite its initial high cost, the Gemini will probably cost less overall than its alternatives (storage batteries, backup system, and some other form of DC or AC conversion device).

In addition, the Gemini will give you higher system efficiency, since no power is lost in storage and since the Gemini itself runs at conversion efficiencies as high as 98 percent.

There are a couple of problems with the Gemini, though. First of all, using it requires that you hook up to utility power. But as we've already seen, the use of wind power is most justifiable in economic terms when your installation is a long way from the power company lines. Thus you can't use the Gemini where you need it most.

The other problem involves the power company. Whenever you produce an excess of power, that excess is routed back through your electric meter and out into the utility grid. Your meter runs backward and this of course reduces your electric bill. Conceivably, you could even come out ahead at the end of the month. You'd produce more power than you used; your meter reading would be lower than it was at the start of the month; and the power company would owe you money. That's where the rub comes in. The power company doesn't like this arrangement. In effect, you're selling them power at retail rates.

So at Windworks, for example, the power company installed a ratchet on the electric meter so it couldn't run backwards. And they installed a second meter to keep track of the amount of power Windworks feeds back into the utility grid. Just how much the utility will have to pay for this power is now in the hands of the Wisconsin Power Commission. This issue, when it is finally settled, will affect the overall efficiency of the Gemini system. If the utility pays you straight retail rates for any power you pump back into their lines, you have in effect, a 100 percent efficient storage system. But if they allow you only 60 percent of retail for the power you produce, your storage system is now only 60 percent efficient.

It all boils down to this: The true picture on the Gemini is not yet in. The future of the system is tied up in the power commissions for each state in the country. Until the power commission in your state comes up with a ruling, you have no way of computing the economics of the system. And your state power commission may not come up with such a ruling for some time. In fact, they may not even be aware of the Gemini system or the need to make a ruling. If you are considering Gemini, you may have to initiate power-commission action yourself.

BUILDING YOUR OWN PLANT

Build your own wind generating plant? There's no reason why you can't if you're handy with tools. There are two ways to go about it: Kits or plans.

Maybe the most sophisticated and complicated plans design is the 12/16 from Helion. The plans show how to build a mill with either a 12- or 16-foot diameter propeller. Outputs are about 1,300 watts maximum for the 12 footer, and 1,500 watts for the 16. Rotor blades are of hollow aluminum construction, and fabrication techniques include heli-arc welding, heat treating, and so on. You won't want to tackle this design unless you know your stuff, but you can also buy the same design in a finished mill under the Kedco 1200 name. Solar Wind Division of Enertech has a unit available in kit or finished form.

All the above mills are conventional horizontal axis designs, but there are some simpler designs available if you are interested in vertical

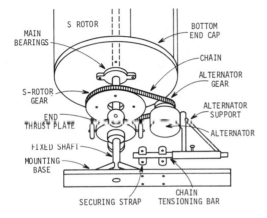

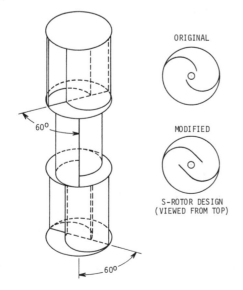

Savonius-type rotor, above, is one of the easiest to build, using 55-gallon drums for the rotor elements. While the Savonius is primarily a low-speed, high-torque type of rotor most suited to pumping, it can be used to drive a generator if geared up. Stacking oil drum rotors, as shown below, is one way to increase power from the unit.

axis mills such as the Savonius. *Wind and Windspinners* by Hackleman and House gives a fairly complete treatment of Savonius generating equipment made from 55-gallon drums. Brace Research Institute and VITA also have plans for oil-drum Savonius designs, but these are meant to pump water, not to generate electricity.

Windworks offers plans for an ingenious tower built from electrical conduit. You might do well to investigate the tower even if you plan on another mill.

How much can you save building your own? Probably more than half of what you would have to pay for a commercial unit in the same power class. But you shouldn't expect the homemade mill to hold up as well as the good commercial designs such as Elektro or Dunlite. These mills are designed to run for years with almost no maintenance. Most homemade designs won't approach that kind of performance; they usually make use of automotive alternators, rubber belt drive trains, and other easy to find, easy to handle components that just can't match the durability of special heavy-duty, slow-speed generators and oil-bathed gearing systems. Still, if you're skilled enough to build your own, you'll be skilled enough to maintain it.

One last note: If you do decide to build your own, you may as well make it worth your bother and build a unit that will produce at least 1,000 watts. I can't see the justification for spending a great deal of time and

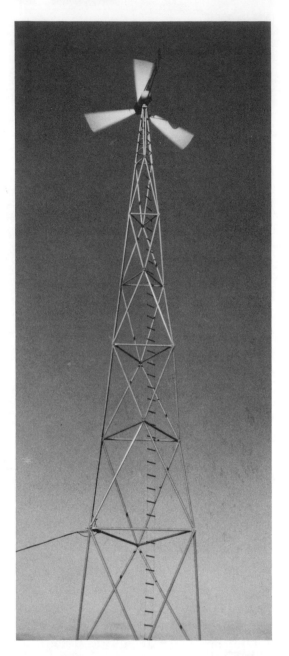

⌐ Reconditioning an old windplant is one of the best ways to get cheap power from the wind. Here's an old Jacobs rebuilt by Windworks, standing atop one of the Windworks octahedron towers. This tower is a good design for do-it-yourself construction. Windworks sells plans for it.

money to build a little generator that puts out 400 watts. At peak output it will do no more than light four 100-watt bulbs.

BUYING A USED WINDPLANT AND REPLACEMENT PARTS

It is conceivable, though not very likely, that you'll be able to pick up an old, used wind generating plant by scouring over the backroads in farm country. Trouble is, this idea has recently become quite popular, so the remaining used windplants are getting very rare, especially those in good condition. The best old windplant you can find is the Jacobs. It may be the best ever built. It was still being made up until 1956, and several models were available, ranging up to a 3,000 watter.

The other windplant you might run across is an old Wincharger. This was a cheaper, lighter unit than the Jacobs, and as a result, a better seller. If you do find an old windplant it's likely to be a Wincharger. The biggest ever made was rated at 1,200 watts.

If you do manage to come up with either a Jacobs or a Wincharger, it's probably going to need some work. With luck on your side, your source may still have his owners manual. Even if he does, you'll probably have to study up on the subject before you can put your rig back in working order. When you do find a used mill, be sure to get as much of the accessory equipment that goes with it as possible. Ask for the control boxes, and the pivoting mechanism on which the whole generator swivels in order to face the wind.

Most likely the wooden propeller will need replacement. You can make your own from wood (Sitka spruce of aircraft quality is the best) or you can try the paper honeycomb and fiberglass technique outlined in the November '72 issue of *Popular Science.* Your best bet, however, is to check out the replacement rotor blades made by Aeropower. Jacobs windplants were very carefully engineered with a propeller design that perfectly matched the generator it was designed to turn. Replace that prop with one of just any design and you'll end up losing watts. So buying a replacement prop will probably pay off in the long run.

To get a quote on the prop you need, round up all the information you can on the windplant you want to restore. Send a description, name plate data, even a photo or two to Prairie Sun and Wind Company. They'll get back to you with a cost estimate. They also sell standard wooden blades in both two- and three-blade configurations. Blades come with stainless steel leading edges, and with hubs. Prices run about $10 per foot of rotor diameter for the two-blade rotors. Three-bladers are about $15 a foot. (Addresses on page 165.)

If you need brushes for an old Wincharger, you can have them custom made on special order, at a cost of 10 for $35. Take one of the old brushes and send it to Becker Brothers Carbon Co., 3450 S. Laramie Ave., Cicero, IL 60650. They'll need the old brush as a pattern for your replacements.

WATERPUMPING WINDMILLS

You can still buy those multi-blade waterpumping windmills that—along with the advent of barbed wire—allowed settlement of the West so many years ago. Over the years, cheap REA electricity forced the windmill out of the picture, and as old mills began to break down, they were simply replaced with electric pumps.

Take a look at the costs for a complete windmill and you can see clearly why they are not your answer unless you need to move a great deal of water. The smallest mill you can buy will cost about $500. Add to that the cost of a 40-foot tower ($1,000), a pump ($130), a cylinder ($50), plus pipe and steel rod. All told, you're talking about an investment close to $1,700. That's a lot of money. But the equipment it buys you is rugged stuff that should last a lifetime without much attention other than a yearly oil change.

Sizing a Windmill. This is a simple operation. Start by estimating your daily needs for water according to the figures in the table on the next page. Multiply the total by five. This is to compensate for the fact that your mill will only pump at full capacity for an

PUMPING CAPACITY

Diameter of Cylinder (Inches)	Capacity per Hour, Gallons		Total Elevation in Feet					
			SIZE OF AERMOTOR					
	6 Ft	8-16 Ft	6 Ft	8 Ft	10 Ft	12 Ft	14 Ft	16 Ft
1¾	105	150	130	185	280	420	600	1,000
1⅞	125	180	120	175	260	390	560	920
2	130	190	95	140	215	320	460	750
2¼	180	260	77	112	170	250	360	590
2½	225	325	65	94	140	210	300	490
2¾	265	385	56	80	120	180	260	425
3	320	470	47	68	100	155	220	360
3¼	—	550	—	—	88	130	185	305
3½	440	640	35	50	76	115	160	265
3¾	—	730	—	—	65	98	143	230
4	570	830	27	39	58	86	125	200
4¼	—	940	—	—	51	76	110	180
4½	725	1,050	21	30	46	68	98	160
4¾	—	1,170	—	—	—	61	88	140
5	900	1,300	17	25	37	55	80	130
5¾	—	1,700	—	—	—	40	60	100
6	—	1,875	—	17	25	38	55	85
7	—	2,550	—	—	19	28	41	65
8	—	3,300	—	—	14	22	31	50

These two tables give you the information you need to order a water-pumping windmill of the right size. See the text heading, "Sizing a Windmill," beginning on the previous page.

AVERAGE WATER NEEDS

Type	Gallons
Milking cow, per day	35
Dry cow or steer, per day	15
Horse, per day	12
Hog, per day	4
Sheep, per day	2
Chickens, per 100, per day	6
Bath tub, each filling	35
Shower, each time used	25 - 60
Lavatory, each time used	1 - 2
Flush toilet, each filling	2 - 7
Kitchen sink, per day	20
Automatic washer, each filling	30 - 50
Dishwasher	10 - 20
Water Softener	up to 150
¾-inch hose, per hour	300
Other uses, per person per day	25

average of four or five hours per day.

Next turn to the "Pumping Capacity" table and find the cylinder diameter required to produce that volume of water. Next, measure the elevation from your well's lowest water level during the dry season, to the level of your pump discharge. Finally, select the rotor size required to handle the cylinder size you have selected at the elevation you have measured.

WINDPLANT TOWERS

Just be sure you get a tower high enough to place your windmill rotor at least 15 feet above all obstructions within a radius of 400 feet.

One of the cheapest and easiest ways to get the maximum output from any wind installation is to invest in a good, tall tower. We've already noted that the power available in a wind is proportional to the cube of the windspeed. Anything you can do to increase the windspeed through your rotor blades will do a lot to increase mill output. That's where the tower comes in. Even in a flat area without nearby obstructions such as trees or buildings, windspeed is anywhere from 25 to 50 percent higher 30 feet above the ground than it is at ground level. Even a 30 percent increase in windspeed can do a lot for mill output. Example: Say your generator will put out 500 watts in a 10 mph wind. In a 13 mph wind you'll get almost 1,100 watts.

So when you consider the fact that a tower can easily double the output of your wind generator, the tower's importance is great. Any tower should be at least 30 feet tall. Beyond that, it should rise at least 15 feet above any object within a radius of 400 feet.

If you decide to buy a tower, you have a choice of two basic types. The cheapest solution is to use the guyed tower type. This is essentially the same type of tower used for radio transmission antennas. It stands on a small, one-legged base and gets most of its stability from guy wires anchored to the ground. Enertech sells guyed towers made by Rohn. Prices are very low compared to your other alternative in prefab towers, the three- or four-legged tower. A 30-foot guyed tower without top section may cost under $400 for a lightweight model, less than half the price of a standard four-post tower. You can get a 60 footer for under $700, just about the same price you'd have to pay for a 21-foot four-post tower.

The three- or four-post tower is the one you've probably seen holding up the water pumping mills on American farms. It stands on a sturdy base, and therefore needs no guy wires. This sturdy construction requires a lot of materials, and that's why these towers cost more than guyed towers. If you live in farm country, you can probably find an old tower still standing. You may be able to pick it up for a good price. But before you buy, make sure the tower is in good condition, and that it is bolted — not riveted — together. You'll have to disassemble the mill tower to take it down and haul it away. This isn't much fun when things are riveted together. With bolted construction, disassembly and later reassembly at your wind site are much easier.

There's one other possibility you might check out in the used tower subject. Those gas station signs you see along the interstates often sit atop very high tubular steel poles. A little scrounging on your part might turn one of these poles up at a good price. Smokey Yunick used a pair of them to mount his experimental wind generator high in the sky above his Daytona Beach garage. These poles are strong, and present a very neat appearance compared to the other types of towers now available.

If you want to try making your own tower, contact Windworks about its octahedron tower. You can bolt it together from ordinary electrical con-

Two basic tower types are illustrated in this photo of Henry Clews' installation. In the foreground a three-legged tower topped off with a Dunlite model 2-kW. In the background is a guyed tower, mounting an Elektro 6kW. Guyed towers are cheaper than three- or four-legged towers.

duit. That might seem like a flimsy material for a windmill tower, but don't be fooled. The tower gets its strength from its octahedron design. A system of triangles gives the construction surprising strength.

Whatever type of tower you choose, one thing is certain. You'll have to anchor it firmly to the ground. A tower will act like a lever in the wind, so the forces trying to pull it free of the earth can be tremendous. Any tower should be set into concrete footings extending four or five feet below the earth. Guy wires for guyed towers will also require concrete anchors set well into the earth.

ROTOR TYPES

When most of us think of windmills, we think of the type with propellers that turn rotors on a horizontal axis. These include the old water-pumping mills so common on farms in past years, and the picturesque Dutch mills used to pump water and grind grain.

But there's an entirely different class of windmills that spin on a vertical axis. Some of these vertical-axis rotors are of ancient design, while others are recent innovations. Let's take a look at some of the various rotors and examine their advantages and disadvantages.

HORIZONTAL-AXIS ROTORS

The best of these will give you the most power possible for a given windspeed. That's the main factor in favor of horizontal-axis designs. They also have one important drawback. In order to function, the rotor must face directly into the wind. Since wind direction is constantly changing, the horizontal-axis rotor will have to be equipped to follow it. This is usually accomplished by fitting a tail vane on the mill that applies wind force perpendicular to the axis of the rotor. Another technique is to position the rotor downwind from its support tower and to cone the blades, or sweep them, so their tips are downwind from the hub.

In any case, a spinning rotor is highly resistant to any changes in its orientation. The rotor acts just like a gyroscope. If it's spinning at a good clip and facing into a west wind, for example, it will resist when a gust comes in from the north and pushes on the tail vane, forcing the rotor to face north. In effect you have the tail vane fighting the gyroscopic forces in the rotor, and the result is a strain on the rotor blades and other parts of the structure. In addition, while the rotor is resisting orders from the tail vane, it is also facing the power of the wind only indirectly. And of course this cuts efficiency somewhat. Still, the horizontal-axis rotor is more efficient than any other type.

Types of Horizontal Axis Rotors. *Modern high-speed rotor.* This rotor is the most efficient design available, and the one most suited to generating electricity. It is designed to spin very fast, and can provide tip speed ratios (speed of rotor tips divided by wind speed) of around seven. This high speed operation is best for generating electricity because electrical generators must turn at relatively high speeds. With a high-speed rotor there is less need to step up rotor shaft speeds with gears or belts or chains. And the less you have to deal with gears and belts or chains, the less power you'll lose to friction.

In order to achieve good high-speed operation, this rotor has two or three slim blades, a lot like those of an airplane propeller. These slim

Kedco 1200, shown here without cowling, is a twelve-footer rated at 1,200 watts maximum output. Blades here are in their maximum pitch position—the position they assume to prevent overrevving at high windspeeds. A system of weights and springs automatically feathers the blades to suit windspeeds. Similar systems are used on other quality windplants. Some of the cheaper versions use a centrifugal air brake to limit revs. Kedco plans to bring out a 16-foot version of this plant.

The bicycle-wheel wind turbine is a new approach to windmill rotor design developed by Tom Chalk. Clark Y airfoil blades of aluminum clip over wire spokes of this 15-foot wheel. Advantages of the design are low weight and cost. Unlike most windmill rotors, however, this one has no provision for handling very high windspeeds. Since high winds can overrev a windplant, and destroy its rotor or generator, all good designs feature a blade-feathering mechanism to spill air at high speeds and limit the rpms of the mill. This Chalk design can't do this, so a clutch of some kind between the rotor and generator might be required to protect the generator. The inherent strength of the rotor makes it immune to damage from over-revving. At present, windplants with Chalk rotors are under development.

rotor blades provide almost no torque at low wind speeds. In fact, most wind generating systems won't even start to spin in winds under seven mph. This lack of low-speed power makes the high-speed rotor useless for operations such as pumping.

Multi-vane pumping rotor. The design of this rotor is directly opposite that of the high-speed rotor just discussed. A waterpumping mill needs lots of low-speed torque to overcome the resistance of the pumping mechanism. The multi-vane rotor provides that torque by presenting a great deal of surface area to the wind. A large number of wide vanes can catch much of the power available in the wind, and ploddingly put it to work running a low-speed pump. But all those vanes just get in their own way and cause drag that prevents the rotor from ever reaching high rpm. The result is a tip speed ratio around two, and low overall efficiency in high winds. But just as the high-speed prop of the wind generating system can get along fine even without low-speed power, the multi-vane rotor does its job well, even without high-speed power.

Sail-type rotors. These are generally best suited to slow-speed applications such as pumping and running mechanical equipment. They fill the area between high-speed rotors and the multi-vane type. Some are, however, fairly close to high-speed rotors in terms of their ability to drive electrical generators. The Princeton sailwing developed by Tom Sweeney is the most promising high-speed sailwing rotor.

VERTICAL-AXIS ROTORS

Vertical axis rotors have two main advantages over horizontal-axis rotors. They can accept the wind from any direction without the need to pivot and face it. And since their axis is vertical, transmission of power down the tower to the ground is simple and direct. You can simply extend the rotor shaft straight down the center of the tower without the need for any complications such as right-angle gearing. You can also locate a heavy or bulky generator at or near ground level instead of high atop the tower. This makes assembly and maintenance simple. Horizontal-axis generating mills usually place the generator atop the tower, and slip-ring commutators are required to route the current down to the ground.

Offsetting these advantages? Most vertical-axis mills are low-speed, low-efficiency designs, best suited to pumping instead of generating electricity. There are exceptions as we'll see shortly.

Types of Vertical-axis Rotors. The most common vertical-axis rotor is the Savonius, developed by S. J. Savonius during the 1920s. This is a simple rotor to build, using cheap scrap materials, and it is ideally suited to pumping applications. This, however, makes it more useful in underdeveloped countries where irrigation

The Darrieus-type vertical-axis rotor shows good promise as a wind generator. This one, developed by Sandia Labs, has Savonius S-rotors inside buckets mounted at the top and bottom to provide starting power. A Darrieus rotor will not start spinning without extra help. Dominion Aluminum Fabricators, Ltd., of Canada is now offering a Darrieus wind system in two sizes, a 20 footer and a 15.

by G. J. M. Darrieus, of France. It has all the advantages of vertical axis design, yet it can nearly match the modern high-speed rotor in performance. The blades of this "eggbeater" rotor form a curve called a troposkien, which is the natural curve a flexible rope or cable will assume if its ends are tied to the ends of a spinning shaft. Since the blades are preformed to this shape, there are no bending stresses on the blades at high speed, so the blades can be made lighter, and more cheaply.

The main drawback of the Darrieus is that it isn't self-starting. It needs outside power to start it spinning. Once it starts, the wind takes over. There is a simple solution to the problem, however. Combine the Darrieus with a couple of Savonius rotors (which are self-starting) on the same shaft. The Savonius rotors start things going; then the Darrieus rotor takes over to provide good high-speed performance.

is a problem, than in the U.S. where electrical generation is the main goal. True, Savonius rotors have been used to drive electrical generators, but their low tip-speed ratios of 1 to 1½ do not lend themselves to generation on a useful scale. (See the drawing on page 171.)

Darrieus rotors. This is a very promising design developed in 1925

WINDPOWER AND THE FUTURE

Take a look at the table of available windpower equipment shown near the beginning of this chapter and one thing is evident: The larger a windplant, the lower its cost per watt. The future of windpower probably lies in large-scale windplants, simply because they will be the most economical. But if the ERDA-NASA windplant in Ohio is any indication of the future, we'll probably have a long

SMITH-PUTNAM WIND TURBINE

NASA WINDPLANT

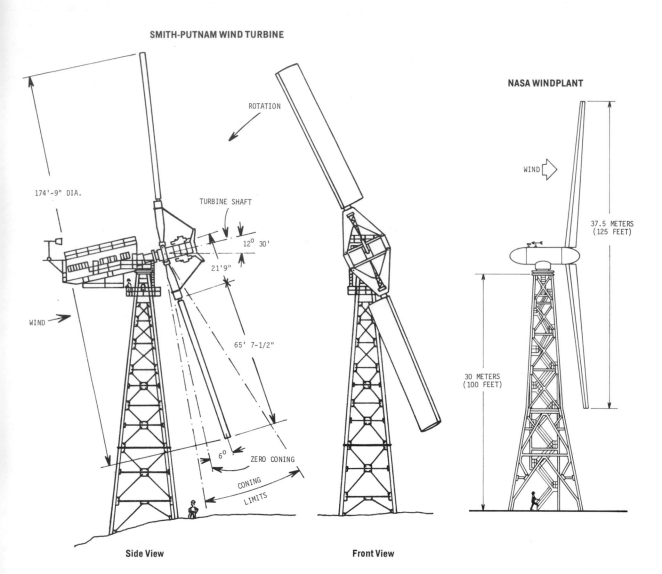

ROTATION

174'-9" DIA.

WIND

TURBINE SHAFT

12° 30'

21'9"

65' 7-1/2"

6°

ZERO CONING

CONING

LIMITS

WIND

37.5 METERS
(125 FEET)

30 METERS
(100 FEET)

Side View

Front View

America's first attempt at large-scale wind generation was the Smith-Putnam Wind Turbine, located atop Grandpa's Knob near Rutland, Vermont. The giant windplant had a two-bladed propeller, mounted downwind from the tower, with the blades swept back or "coned" to allow the mill to face into the wind. The mill met an early end when vibration and stresses fatigued one of the metal blades and tore it off. Vibration problems are also plaguing the new NASA windplant near Sandusky, Ohio. This new plant also has a two-bladed prop mounted downwind from the tower, but this design may be responsible for many of the vibration problems. Located downwind from the tower, the prop passes through the tower's "wind shadow." This causes uneven running. A two-bladed prop also creates problems when it swings to follow wind direction. This is because the blades have a strong gyroscopic resistance to being pivoted about the tower when they are in a horizontal position. But when the blades pass through a vertical position, they have no resistance to pivoting. A three-bladed prop solves this problem, but costs more to build.

wait before windpower contributes much energy nationwide. This giant research mill is rated at 100 kilowatts, and should be able to supply the power for about 30 homes. Unfortunately, the plant is plagued with problems. And it's not alone. All the other large-scale attempts at windpower have failed, too. India, Russia, the U.S., and Germany have tried, and none have succeeded—for one reason or another. The main problem is that as mills get bigger, strength and weight become more critical. Cuts in weight improve performance and economics, but decrease strength and result in breakdowns. If these problems can be solved, the way to large-scale windpower may open up.

In the meantime, what is the future of home-size windplants? The main drawback with the windplant of today is the storage system. It's expensive, and a poor use of resources. If every home in the U.S. had to rely on a bank of lead-acid storage batteries, the world supply of lead would be under tremendous pressure. This, of course, would raise the price of storage systems even more.

Something like the Gemini inverter by Windworks could save the day, but only if it could be mass produced to lower costs. And this can result only if good agreements can be worked out between the utility companies and the consumers. Hans Meyer of Windworks believes that a mass-produced windplant and inverter system could produce electricity for three or four cents a kilowatt hour. This would make such a system competitive with utility power in large areas of the U.S. But all this assumes cooperation from utilities.

This is what that cooperation would entail. First, the utility would have to allow anyone to hook his wind system up to the utility power grid. The utility would have to consider each home thus hooked up as part customer, part supplier to the utility. The utility would have to pay as much for the electricity you supply them as you do for the electricity they supply you (or close to it). If the utility pays you substantially less, your windplant is no longer going to produce at three or four cents a kilowatt hour. Your costs will rise.

In short, for this system to work, the utility will become more a cooperative than a corporation. Probably nothing short of government legislation is likely to bring about that kind of change.

HEATING WITH WOOD

Chances are you can heat your home with wood for less than you now spend to heat with oil or electricity. There's a lot of wood around to be burned, and it burns clean. Burning it can even be good for our forests. And unlike coal, oil, and gas, wood is a renewable resource.

Burning wood for heat lacks the glamor of solar and wind power, but for most people wood power makes more sense. It's easier and cheaper to install, and it pays for itself faster. If you have your own woodlot and a chain saw you could buy a good woodstove for the price of a heating season's worth of oil or electricity, then heat your home virtually for free. Even if you have to buy wood, if

you shop wisely you'll still save money over conventional heating fuels. Of course, heating with wood is more work, but probably not as much work as you think. Modern wood stoves are remarkably efficient—you have to fuel them only twice a day. You can even buy combination wood/gas or wood/oil furnaces that will automatically heat your home with gas or oil if you forget to refuel with wood, or if you're away from home for extended periods. Ash removal? It's about a twice-a-month affair. And those ashes are good for your garden.

Even if you have no desire to use wood as a main source of fuel for home heating, you should seriously consider it as an auxiliary source.

Burning wood can easily cut your conventional fuel costs in half. And wood heat is an ideal backup for solar heating systems.

Availability of Wood. Despite what you may think, there's a lot of wood around to be burned. The total amount of wood growing in the U.S. is actually increasing every year. There are a lot of reasons for this, including modern replacements for wood in many commercial products, increased growth due to the extra carbon dioxide man has been putting into the atmosphere, and the reversion of many Eastern farms to forest.

At any rate, the U.S. Forest Service has estimated that every year there are some 13 billion cubic feet of fuelwood produced every year in the U.S. This is enough to heat just about every home in the country. And remember: This is wood that is left over after natural mortality and commercial timber cutting. And surprisingly, most of this wood is available in the East, where population—and consumption—are highest.

Burning wood is a good way to get rid of what otherwise might be wasted. Logging and manufacturing wastes make good fuel. Manufacturing wastes are often in the form of sawdust or chips, but there are woodstove designs for burning just such wastes. Logging wastes—tops of trees and branches—are often left to rot. Using them to heat homes makes more sense. Cull trees, so twisted or stunted that they have no timber

value, still make fine fuel. So do diseased trees, or those downed by wind. And burning any of these forms of waste wood saves our rapidly dwindling supplies of fossil fuels. Removal of cull trees and those under attack by disease also opens up the forests and provides more space for the healthy specimens. This in turn increases the output of quality lumber.

Of course cutting wood is not entirely without bad side effects. While the removal of old and dead trees may help the forest grow more efficiently, and may even bring about an overall increase in the number of animals living in the woods, it does harm some wildlife. The ivory billed woodpecker is one example. If not extinct right now, the ivory bill is very close to it. Why? This bird has an unbreakable habit of feeding on bugs that live under the bark of long-dead trees—old decaying trees with the bark peeling off in big slabs. These are the trees of a terminal forest, and very few exist today. Heavy logging has harvested them long before they got a chance to age.

Against this you can balance the fact that logging has benefited species such as the whitetail deer. There are more deer alive today than when the Indians had the country to themselves.

A Clean Burn. But doesn't wood fill the air with smoke and pollutants as it burns? Yes and no. As Larry Gay explains in his fine book *Heating* *with Wood*: "The products of combustion of wood . . . would be liberated in the forest by decay anyway and do not, therefore, lead to a net increase in environmental pollution."

Buckminster Fuller once described the burning of wood as "the sunlight unwinding from the log." And that's exactly what it is — the reverse of photosynthesis. When wood burns it returns to the air the major components that formed the log in the first place: water, carbon dioxide, and energy. The only difference is that the energy is in the form of heat instead of sunlight.

Of course wood doesn't always burn completely. Oxidation is rarely, if ever, carried to the ideal extreme. The result is a variety of harmful pollutants including carbon monoxide and certain acids. But again, these compounds are very much like those formed in the forest by decay.

Photochemical smog is not solely a product of the Industrial Revolution. Forests have been producing it for millions of years. The products of forest decay can often form a haze. As Larry Gay points out, this haze is "so pronounced in the Smoky Mountains that they take their name from it."

WOODSTOVES, FIREPLACES, FURNACES: WHAT TO BUY

To get any heat out of wood, you have to burn it. The amount of heat you get is determined largely by what you burn the wood in — fireplace, stove, or furnace. The most efficient wood burners are the airtight stoves and furnaces. The good ones can be about 60 percent efficient, but usually come in around 50 percent. This means they can put about half of the heat produced by burning wood into your home. After airtights, ordinary stoves are next in efficiency, followed by openfaced stoves like the "Franklin," then by circulating fireplaces, freestanding fireplaces, and ordinary open-masonry fireplaces. The latter are terribly inefficient — in some cases negatively so. They can actually lose more heat than they put into your home. If you seriously want to heat your home with wood don't even consider an open-masonry fireplace. But if you already have one, there are ways to make it perform better, as we'll see later.

Airtight Stoves and Furnaces.   These should always be your first choice for efficient home heating, so let's look at them first.

The secret to efficient burning of wood is control. You have to be able to control the fire to make it burn slowly, and to make it burn the wood as completely as possible. Airtight woodburners give you the control you need. Other types of woodburners do not. To understand the operation of an efficient stove, you must first understand some of the basic principles of wood combustion.

Wood is not strictly a single fuel,

but a mixture of fuels: solids, liquids and gases. And wood burns in three distinct phases which may proceed in sequence or all at once. In the first phase, free water is driven out of the wood, mostly be vaporization caused by the heat of the fire. The wetter the wood the more heat is wasted boiling off the water. That's why you'll get more usable heat from dried or seasoned food.

The heat from phases two and three is what powers phase one. In phase two, the wood begins to break down into charcoal, gas, and volatile liquids. As much as 50 percent of the potential heat inside a log is released by the burning of the gas and volatile liquids—if the woodburner is designed to allow complete combustion of these products. Yet many woodburners allow these gases and volatiles to escape—unburned—up the chimney.

In phase three, combustion of the charcoal takes place. It is the charcoal-burning phase that provides most of the useful heat. Now let's move inside a stove and see how all this affects efficiency.

About an hour ago you filled the stove with a good load of wood. The fire has had a good chance to get established and it's roaring along at a good clip. Too good a clip! The stove is overheating, pouring the room full of excess heat. The wood is being consumed so fast it will be nothing but ashes in a couple hours. Flames are licking way up the stove pipe and it's glowing red hot.

SIMPLE BOX STOVE

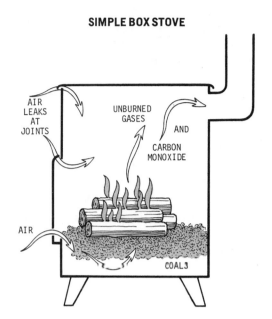

An ordinary box stove has no provision for secondary air and lets volatiles pass directly out the stack as they rise from the fuel load. Efficiency is low (about 30 percent). Slow-burning, long-lasting fires are difficult to achieve.

To slow the fire you close down on the draft. If your stove is airtight, the fire will slow down almost immediately. If it's not airtight the fire will continue to rage, then slowly start to subside. But it will not drop to as slow a rate as the fire in an airtight, because air leaking into the stove around doors, burners, and other joints provides enough oxygen for rapid burning.

What does this mean? The leaky stove will most likely produce too much heat over a very short period of time, then burn out. Meanwhile the airtight can burn along at a controlled rate, producing just the right amount of heat for a period of up to 14 hours.

The airtight keeps you comfortable for half a day; the other stove makes you too hot for a few hours, then lets you get too cold.

That's the big advantage airtight stoves and furnaces have over all other types of woodburners. They allow you to control the draft and thus control the fire. More than that, they let you control the draft at the intake. This is a far better system than controlling draft at the exhaust with a damper in stovepipe or chimney. Dampers are not as positive in their control as draft intake controls. And if you close a damper too far you get smoke in your house.

Some stoves—most notably the Ashleys and Riteways—even have automatic intake controls. You set a thermostat to the desired level and that's it. If the fire starts to get too hot, a bimetal coil closes the air intake to slow things down. When the fire gets too cool the intake opens, extra oxygen reaches the fire, and the rate of burn picks up. Other airtight stoves—the Scandinavian designs—get by without thermostats, as we'll see later.

So control is the first key to efficient woodburning. Without it you get uneven heat over a short period of time, and you waste fuel. With control you get even heat over a long period of time from a single load of wood. And controlled draft means more than just slow even burning; it limits the volume of warm room air that can escape up the chimney. The big weakness of any non-airtight woodburner—and especially fireplaces—is that they allow warm room air to flow up the chimney in torrents. What replaces that air? Cold air from outside your home. You then have to burn more wood to heat that air, and then it escapes up the chimney and so on.

Complete Combustion. This is the next thing to look for in a woodburner. Any combustible component of a log that passes out of the firebox unburned is costing you efficiency. The more completely a stove converts wood to oxides and ashes, the more heat you get out of every stick of wood.

Research has shown that a two-inch bed of coals will consume all the oxygen in the air admitted into a stove under normal draft conditions. In this case, almost all of the carbon is being oxidized fully into carbon dioxide. This is ideal. But if the bed of coals is thicker—say three inches—the ideal of complete combustion is lost. Now a very large amount of carbon monoxide is produced. The carbon monoxide rises up out of the bed of coals, and unless extra oxygen is present above the fuel bed, this half-burned carbon will pass up the stove pipe. Half-burned carbon means less heat produced, less efficiency, and a waste of wood.

So we see that it's critical to get some oxygen into the space above the fuel bed where it can further oxidize the carbon monoxide to carbon dioxide. In practice this is done by introducing "secondary air" above the fuel bed. This air, brought in through a secondary air intake, does more than oxidize the carbon monoxide. Remember phase two? Gases and volatile liquids are being driven out of the wood by the heat of the fire. These gases and volatiles must also be supplied with oxygen if they are to burn. The secondary air takes care of that job, too. Since we already know that about half the heat we get out of a log can come from burning these gases and volatiles, you can see how important it is to get secondary air into the stove.

As it turns out, about 80 percent of the air introduced into a stove should be supplied over and around the fuel bed. The rest should feed the glowing charcoal.

There's one more reason for admitting secondary air into the stove. In airtight stoves—especially those with thermostats—a lot of volatiles are driven off when the stove is adjusted for a long slow burn. So is a lot of free water. If these volatiles are not burned, they can combine with the water vapor, and cool as they rise up the chimney. There they condense on the chimney walls and form what is known as "creosote." This stuff can cause corrosion; it can drip out through joints in the stove pipe. And next time you open the draft for a quick burst of heat, it can ignite and cause a fire in the chimney. In a good chimney this fire should be no hazard. In an old cracking chimney it can start a housefire. There are some ef-fective ways to curb production of creosote, and we'll cover them later in this chapter in the discussion on safety.

One last fact to consider about secondary air: It should be preheated before it enters the firebox. Volatiles and gases have an ignition temperature of 1,100 degrees F. or higher. If they are allowed to cool below that temperature they will not burn, no matter how much oxygen is present. If secondary air is not heated before it reaches these gaseous fuels, it can cool them below their ignition temperature. They'll escape up the chimney unburned, or worse, condense in the chimney if the flue temperature drops below about 250 degrees.

Rating the Best Units. Okay, that's a quick look at the principles of wood combustion and stove design. Now let's take a look at some of the best stoves and why they rate on top.

Ashleys. There are now well-established favorites. They are relatively inexpensive, they'll burn just about any wood, and they'll hold a fire overnight with ease. They come in two basic models—circulating console styles and the economy line. The console models look like most circulating wood stoves. They have a sheet metal cabinet around a firebox and resemble gas space heaters more than wood stoves. The economy line has a bare-bones wood-stove look. I mention this not out of style consciousness, but only so you'll know one type from the other.

AIRTIGHT STOVE (Ashley)

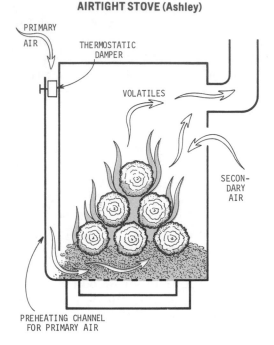

PRIMARY AIR

THERMOSTATIC DAMPER

VOLATILES

SECONDARY AIR

PREHEATING CHANNEL FOR PRIMARY AIR

The Ashley airtight preheats primary air before passing it into the fuel supply. An automatic thermostat assures an even rate of burn. So a load of logs can provide overnight heat without attention. Note the secondary air inlet at back of stove.

The Ashley shown in this home provides the bulk of the heating requirements. This is the economy model. A more expensive model has a metal cabinet around the fire box. The damper for the fireplace behind the stove prevents backpuffing and loss of draft.

In practice, both types are excellent sources of heat. A plant ecologist I know heats his entire home near Ithaca, New York, with an Ashley Columbian. He lives on 50 acres of woods, so aside from chainsaw fuel and maintenance, his heat is free. The same model stove heats the uninsulated home of another friend in northern Vermont. And it keeps it comfortably warm overnight even in subzero weather.

If there's any problem with Ashleys, it's their tendency to produce creosote. This usually happens overnight when a full load of logs is allowed to burn with the thermostatic draft closed down. Only enough primary air is admitted to feed the glowing coals. Volatiles and water are driven out of the load of logs and into the relatively cool chimney where they condense to form creosote. My ecologist friend had one minor chimney fire due to deposits of creosote shortly after he started using his Ashley, but he's learned how to avoid the problem, as covered later in this chapter, under the heading "Wood-burning Safety Tips."

The console-model Ashleys have automatic secondary air intakes that provide for more complete combustion of volatiles. This is a partial solution to creosote formation, but it's not a total success. So Ashley also sells a patented draft equalizer to further reduce the problem. This is a short section of stove pipe that fits between the stove and the main stove pipe. Built into the equalizer are ducts that allow air to enter the stovepipe for further oxidation of volatiles. Since both the draft equalizer and the secondary air inlet help achieve complete combustion, they boost efficiency while fighting the creosote problem.

In all fairness to Ashley, they don't have a corner on the production of creosote. As the owner's manual for the excellent Riteway 2000 woodburning heater puts it: "All types and makes of woodburning equipment will give trouble with creosote deposits under certain conditions." And according to the Jotul catalog: "If you learn to run it (a woodstove) well, you should have plenty of heat and very little creosote." So I wouldn't let Ashley's reputation as a creosote producer scare me off. With prices starting at a bit over $100, Ashleys are one of the best buys in woodheating—if you can find them. As is the case with most woodheaters, they are in very short supply at present.

Stoves similar in design to the Ashleys are made by several other firms, including King Stove and Range, Brown Stove Works, United States

Thermostatically-controlled stoves like this one from Shenandoah can hold a fire for 12 hours or even longer. This extra-large door makes stoking easy.

Stove (also sold through Sears), Locke Stove, Shenandoah Manufacturing, Washington Stove Works, and others. Many of these are more substantially made than the Ashleys, with heavy cast iron parts and firebrick linings. (See the address list at the end of this chapter.)

Riteway. If Ashleys are the best-known stoves, Riteways are a close second. In terms of versatility, they probably deserve to be first. Riteway not only makes a fine woodburning heater, they also make woodburning furnaces, boilers, and even multi-fuel

BASE-BURNING (Riteway)

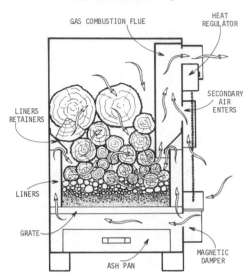

Riteway's base-burning design forces smoke and heated gases back into the fire where they mix with secondary air for secondary combustion, before passing out the flue.

furnaces that burn wood and your choice of oil or gas.

The Riteway 2000 radiant heater is a large, ungainly looking, and boxy heater. If its looks bother you, you can enclose the whole stove in a cabinet. Like the Ashley, the 2000 has a thermostatically controlled draft, but its draft flow and internal structure are more sophisticated. The passage from firebox to flue is located at the base of the firebox instead of at the top. This forces the volatiles to pass down and close to the hot coals at the base of the fire. This heats the volatiles which then pass into a secondary combustion chamber where they are

mixed with secondary air, and where they burn.

Do not buy a stove or furnace that's bigger than you'll need. A stove burns cleaner and more efficiently at half throttle than it does at idle. And by the same token, don't undersize a woodburner or you'll have to feed it wood more often than you'd like. Price of the Riteway 2000 is currently around $300.

For the ultimate in heating versatility, consider Riteway furnaces and boilers. They use the same complete combustion draft pattern as the 2000 stove, but they go several steps beyond. They'll burn both coal and wood. Most models will accept accessory oil or gas burners. You set the thermostat on the accessory burner a few degrees below the thermostat on the wood or coal burner. If the wood or coal fails to maintain the desired temperature (because you forget to stoke the fire or because you aren't home to do so) the gas or oil burner takes over.

On top of all this, Riteway furnaces and boilers will take accessory water heaters and even humidifiers. Like ordinary gas or oil furnaces, the Riteways have warm air circulation blowers, air filters, heat exchangers. Riteway boilers come in both steam and hot water models. Btu capacities run from 125,000 for the LF-20 up to 350,000 for the LB-70. Prices run from $1,342 up to $3,840.

If you're interested in Riteway furnaces or boilers, the Riteway engineering department can help you choose the right one for your needs. You provide them with a rough sketch of your floor plan, indicating room sizes, numbers and sizes of doors and windows, ceiling height, location and size of your chimney, and details on wall construction and insulation.

Installation of one of these furnaces makes the most sense in a new home, or in one that needs a new furnace. My experience has convinced me that a good woodstove will do a surprising job of heating an average-sized home —and it does the job without a lot of complicated ductwork or plumbing (required with a furnace or boiler). And you can get the equivalent of dual-fuel heating with a woodstove and your present heating system. Just set your present system's thermostat to around 60 degrees. Whenever your woodstove lets the house drop below that temperature, your present system will take over. (See the address at the end of this chapter.)

Bellways. Another maker of wood heaters and furnaces is Perley Bell. His equipment sells under the Bellway name. Bellway furnaces will do everything the Riteways will do; heat your home, heat water, humidify, and run on dual fuel (just wood and oil at present). Bell also has information on combining his Bellway with your present oil burner for automatic operation.

Bell tells me he's "spent over 25 years on better woodburning equipment, not dreaming there would ever be such a market as now." Over those 25 years he has developed an ingenious draft system. His furnace is a "base burner" like the Riteway, and as with the Riteway the primary draft is not infinitely adjustable. Acting on orders from an upstairs thermostat, the main draft is either open all the way, or shut tight. Secondary air is controlled automatically by the air pressure within the furnace.

When the primary draft is closed, just enough air to maintain the fire diffuses in through the secondary air port. Since this air reaches the fire by going in what is normally the exhaust port of the firebox, it is impossible for too much air to enter. As soon as the fire rises above the "low idle" stage, the hot gases in the firebox expand and pass out the exhaust port, blocking the infusion of more air.

The end result of all this is that the fire idles along very slowly (burning very little wood) until the thermostat calls for more heat. The primary draft opens and the fire starts to supply heat almost immediately.

Bell's heaters use the same design, but the intake control is a bimetal thermostat like the type used on other 'stat-controlled stoves. Both heaters and furnaces are well made with heavy arc-welded steel bodies and fire brick linings. The furnaces take logs up to four feet long; the stoves will handle two-foot logs. Prices? About $3,000 for a 125,000-Btu furnace, $500 for a 20,000 Btu stove. A larger, 35,000 Btu stove is about $800. (Address at end of chapter.)

Logwood Furnace. Another interesting woodburning furnace is made by the Marathon Heater Company. The big innovation in this furnace is its firing-air tubular grate. With this grate all combustion air is passed through the tubular grates that support the fire. Since these tubes are covered with glowing charcoal, the air is super-heated when it enters the firebox and the secondary combustion or gas-burning tube. Marathon claims this arrangement contributes to more complete wood gas combustion—even at low rates of burn—and consequently to high efficiency and low creosote production.

Marathon's Logwood furnace also has a secondary heat exchanger welded in place atop the furnace. This exchanger provides extra space for the wood gases to burn before they pass out the flue. This is especially important at high burn rates when wood gas flames are very long. Without the exchanger these flames may reach into the chimney where their heat can't be captured and put to use.

The Logwood comes in two models, one developing 200,000 Btu, the other 125,000. Prices are about $1,300 and $1,700. While these prices are lower than those for other furnaces mentioned, they do not include such things as controls, blowers, filters and so on. These will add another $400 or so to the final cost. (Address at end of chapter.)

DownDrafter. If forcing the volatiles to pass close to the bed of coals is good, then passing them down and

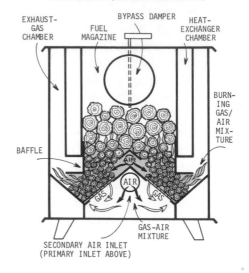

DOWNDRAFT (Vermont Woodstove)

Vermont Woodstove's DownDrafter forces gases down through a coal bed where they are superheated. Then they come into contact with secondary air and are burned. This is the only true down-draft stove.

actually through the coals should be even better. The coals will superheat the volatiles to well above their ignition point (around 1,100 degrees F.). Secondary air, introduced just below the coals, will provide the oxygen for clean burning. The nearby coals will assure ignition of the volatiles. In theory this system is superior to the updraft system; since the gases are mixed with oxygen immediately after passing through the coals they have no chance to cool off and fail to ignite. In an updraft stove the gases have to pass through the whole stove full of wood and consequently have ample opportunity to cool—especially if the stove has just been loaded with cold wood, and especially if the fire has been turned down for a long burn.

That's the theory. The problem is that downdraft stoves are balky. Normally the draft would much rather go up than down. Heat rises. Downdraft stoves aren't new (Ben Franklin made one for coal), but they've never succeeded. Smoking and backpuffing are a problem unless the chimney draws well, and the stove owner tends his fire skilfully. As far as I know, Vermont Woodstove's brand new DownDrafter is the only true downdraft stove presently available. Ashley claims their stoves work on the downdraft principle. But air is merely ducted from an intake near the top of the stove downward through a passage outside the firebox. The air then enters the firebox at the bottom and flows upward through the fire. This downward trip may preheat the air, but it doesn't qualify the Ashley as a downdraft stove.

The DownDrafter seems to have solved the problems inherent in downdraft design, and it should prove to be an efficient stove. It's equipped with a blower that forces air through a space surrounding the firebox, and on out the front of the stove. This helps extract heat from the stove when it's running at full clip. The blower—and the draft—are both controlled by a thermostat. Unlike the stats on other stoves, this one senses flue temperature, not firebox or room temperature. If the blower is put out of operation due to a power failure the stove will continue to run at a slightly reduced efficiency. Price of the DownDrafter is around $460.

The stove is made in a single model producing around 50,000 Btu per hour. (Address at end of chapter.)

Defiant. Here's a brand-new stove that combines both good engineering and good looks in a single package. At first glance the Defiant looks like a typical Franklin. But open it up and look inside and you're in for a surprise. Thermostat, baffles, secondary combustion chamber, damper, air preheat tubes tell you this is no ordinary parlor stove. The Defiant is, in fact, quite unusual.

It features a unique cross-flow draft pattern. When the stove is burning in

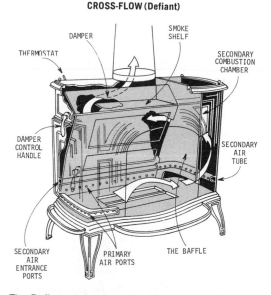

CROSS-FLOW (Defiant)

The Defiant parlor stove is styled like a Franklin, engineered like nothing else. Primary air is controlled by a thermostat, and preheated. Exhaust and wood gases exit the firebox at the right side and pass by the secondary air tube. Then the gases double back to the left, back again to the right, and finally out of the stove. This exposes an extremely large cast-iron surface area to the flame. The result is good heat transfer.

its stove mode, primary air enters the left side and the back of the fire box. This air is preheated before it enters the box. Exhaust gases can only leave the firebox by passing under a large baffle at the right side of the stove. As they exit, they pass right by the secondary air tube, which feeds preheated air into a secondary combustion chamber. This burns the volatiles. But before the exhaust can pass out of the stove it must turn back toward the left side of the stove, and then again back to the right. This long tortuous route results in a 60-inch flame path. And a long flame path keeps the heat in the stove as long as possible, allowing the cast iron to absorb the heat before it can slip up the chimney. Once the iron absorbs the heat it can then radiate it out into your home.

This complex draft pattern makes lighting the stove a bit tricky. But Defiant engineers solved that problem with a damper. Open the damper and the stove can operate in a regular updraft pattern—out of the firebox and up the chimney. Once the fire is established, close the damper and the long flame path is in operation. The damper serves another purpose. For those times when you wish to burn the stove with the doors open, the damper is also left open. The stove will smoke if you try to force the long draft into operation with the doors open. (Address at end of chapter.)

Scandinavian Stoves. Perhaps the simplest, best-made, most attractive stoves available today are the Scan-

dinavian designs. The best known are the Jotul, Trolla, and Lange. All three makes are cast iron. All are available in colored porcelain finishes. All are very efficient with many years of heating Norwegian homes behind them.

These stoves are neither updraft nor downdraft designs. Instead their draft follows an S-shaped path through the firebox. The reason for this is simple: In all other stoves and furnaces mentioned so far, the load of logs burns from the bottom up. The three stages of combustion occur more or less in sequence: drying, liberation of volatiles, burning of charcoal. Left to its own, a fire of this type tends to get hotter as the sequence proceeds. This is prevented with thermostats that cut the draft to keep the heat constant.

But the S-shaped draft pattern in Scandinavian stoves makes thermo-

FRONT-END COMBUSTION (Jotul 118)

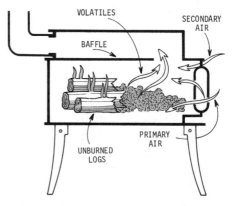

Jotul and other Scandinavian stoves use baffles to encourage a slow, even burn from the front of the logs to the rear. This cigar-like burning pattern prevents the whole fuel load from burning at once and eliminates the need for a thermostat.

stats unnecessary. The draft enters the front of the stove, hits the front ends of the logs, and—with well-seasoned logs—burns them from front to back like cigars. A baffle over the logs guides the draft back toward the front of the stove where secondary air ignites the volatiles. Then the draft passes over the top of the baffle and back towards the rear of the stove and out the flue. The smooth, even, front-to-back burn combines the three combustion stages into one, so the fire doesn't tend to get hotter as it progresses. No thermostat is required. You simply set the draft to a desired level and the stove's burn pattern regulates itself. The baffles not only form the draft pattern, they hold the heat in the stove longer, cutting heat loss up the chimney.

Some of these stoves are quite small, so they steal little space in your home. And many—the Jotul 602 is an example—are so low you can fit them into fireplace openings. Other models include Franklin type fireplaces that are airtight when the doors are closed. This is the only truly efficient way to go if you want to be able to watch the dancing flames of an open fire now and then. Prices for Scandinavian stoves run from about $200 on up to over $700. (Addresses at end of chapter.)

INFERIOR STOVES

If you're truly serious about heating your home with wood, you'll do it with an airtight stove or furnace.

Those mentioned above are not the only ones to consider, but they are some of the best, and the most widely known. Any of them would make a good choice. If you decide to use any form of woodburner other than an airtight you'll be losing efficiency. This will cost you money if you buy wood; time and effort if you cut your own. And it may mean you'll have to resort to other fuels to do a complete job of heating.

On the other hand, you may have a reason or reasons for passing airtights by. Maybe you want the old-style look of those cast-iron stoves from Portland Stove Foundry or Washington Stove Works. Or maybe you want a Franklin (nothing like Ben's original design). Or maybe you've found a good buy on a true antique and you are in love with the old-fashioned styling. If so, go ahead and sacrifice heat for style. But because this is a book on energy, not on interior decorating, I offer no further advice on non-airtights.

FIREPLACES—FREESTANDING AND CONVENTIONAL

If you already have a fireplace, chances are it's actually costing you energy. About 90 percent of the heat produced by the wood it burns goes directly up the chimney. Add to this the fact that a lot of heated room air escapes up the chimney while the fireplace is in use and while it's cooling down. What you have is probably negative efficiency. The best thing to

Better'n Ben's stove is designed to mount over a fireplace opening. It requires no extra flue pipes.

do is close off the fireplace and install an airtight stove in front of it. Or install one of the new woodstoves made specifically to fit over the front of a fireplace. The new Better 'n Ben's, from C & D Industries, is an example. It's an airtight, internally-baffled stove with a large backplate that covers and seals off the fireplace opening. No stovepipe is required since the back of the stove opens up into the fireplace. The stove converts to a freestanding unit if desired by means of a $25 adapter kit. Price of the stove itself is $250. (Addresses at end of chapter.)

If you can't afford the space for a stove in front of an existing fireplace, you can still improve efficiency in other ways. You'll always do best with efficiency boosters that close off the fireplace opening. The Fuego III is a good example of a fireplace improver of this type. It's essentially a metal box that fits inside the fireplace opening and holds the fire. Its front is closed off with tempered glass doors so heated room air can't escape up the chimney. There's space for room air to circulate around the metal firebox. Convection pulls room air in around the base of the box and then passes it —fully heated—back into the room. The Fuego fits flush with the front of your present fireplace and steals no room space at all.

A somewhat similar design is the Stovalator. This unit does not fit flush with your present fireplace, so it steals about a foot of room space. But it presents more surface area to your room, so it may radiate a bit more heat than the Fuego. Neither the Stovalator nor the Fuego can be expected to provide as much heat as a good freestanding stove, however.

Fireplace efficiency boosters that leave the front of the fireplace open are only marginally effective in my experience. This includes the tubular grate types. These consist of fireplace grates with tubular members shaped like the letter "C." The fire rests inside the "C," heating the tubes. Room air flows into the bottom of each tube, is heated, and passes out the top of the tube and into your room. Makers claim these units will double the amount of heat you get from a fireplace. This may be, but you get so little heat from a fireplace that doubling it doesn't help much. I ran one of these units in my fireplace and couldn't detect any noticeable improvement in heating over the unaided fireplace. This particular unit relied on convection entirely. There are others with blowers that should improve performance to some degree. But any device that leaves the fireplace opening free to pass room air up the chimney is never going to do a satisfactory job of heating a home. Prices of even the convectional units start in the same neighborhood as some of the cheapest airtight stoves. And of course, blower-equipped grates will cost even more. Clearly there are better ways to spend your wood-heating money.

What about warm-air circulating fireplaces, those that work much like the Fuego and Stovalator units described above? If you have one and you're happy with it, good. But I certainly wouldn't recommend installing one as a serious source of heat. True, they're a vast improvement over conventional fireplaces. But they can't match the performance of a good woodstove. They cost more. And they require a more elaborate, time-consuming installation. Consider them only if you're more concerned with attractive styling than you are with heat.

This same advice applies to free-standing fireplaces. You can't beat

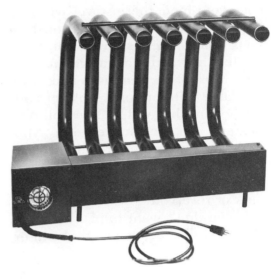

Fireplace grates with blowers to force air through tubular grate members can boost fireplace efficiency, but performance will still not equal that of a good woodstove. And a grate can cost as much as a stove.

them for modern good looks, but you can for heating efficiency. The only one worth considering (that I know of) is made in Norway by Jotul. It has sliding metal doors that close off the fireplace, effectively turning it into a woodstove. According to tests made by Jotul, this fireplace has an efficiency very close to that of some of their fine airtight box stoves.

COOKING WITH WOOD

Wood ranges for the kitchen are still being made by Washington Stove Works, Portland Stove Foundry, Autocrat, Jotul, Lange, and others. Besides providing an economical way to cook your food, these stoves will go a long ways toward helping you heat your home. This is an advantage in winter, but a disadvantage in summer. Most of the American makes are available with optional tanks or coils for heating hot water. The coil-type heaters can be hooked into your plumbing system to provide running hot water. This is a more convenient arrangement than the stove-mounted tank or reservoir.

If you don't want to plan on wood cooking full time, you might be interested in the circulating kitchen heater. This is like a very narrow wood range with a body that doubles as a circulating-type wood heater. The Autocrat 3666 is an example of this type of heater. It's less than 15 inches wide, has a cooking top about 18 by 11 inches, and can be fitted with water heating coils. You can heat and cook with a rig like this during the winter, then change over to a conventional range for summer cooking when the excess heat of a wood stove would be oppressive.

Don't forget that many stoves designed primarily for heating have flat tops that let you use them for cooking, too. Plus, you can keep a kettle of water boiling for humidification and for dish and tea water. Other stoves — like the Jotul Lumberjack No. 380 — are designed for the backwoods cabin where they double as heaters and cookers. This particular stove combines the Jotul front-end combustion system with a two-burner cooking top. (Addresses at end of chapter.)

STOVE KITS—BUILDING YOUR OWN

Kits for turning 55-gallon drums into woodstoves are available from Portland Stove Foundry, Washington Stove Works, and other sources. These kits consist of cast-iron legs, flue collar, and loading doors with draft control. Prices are around $50. After cutting the necessary openings for door and flue collar, you fasten the kit parts to the drum and the stove is ready to install. Since 55-gallon drums are quite large, you get a stove with great capacity. Stoves of this type are so popular in northwest Canada they are called Yukon stoves. (Addresses at end of chapter.)

You can improve on the efficiency of these stoves by incorporating ideas from commercially made wood burners. One idea is to build a 30-gallon drum into the larger 55-gallon drum to achieve a Jotul-type draft pattern. To do this, just place a few inches of sand in the bottom of the larger drum, place the 30-gallon drum inside on the bed of sand, and bolt the drums together through the rear to prevent shifting. This arrangement eliminates the need for welding. Drum stove kits I've seen have no provision for secondary air inlets above the primary air, so you'll have to improvise your own if you want to try for complete combustion. Most stove makers sell replacement parts for their stoves. So you can buy an adjustable draft control and add it as a secondary air intake.

And since Ashley and others sell replacement thermostat controls, you

BARREL STOVE WITH SCANDINAVIAN DRAFT PATTERN

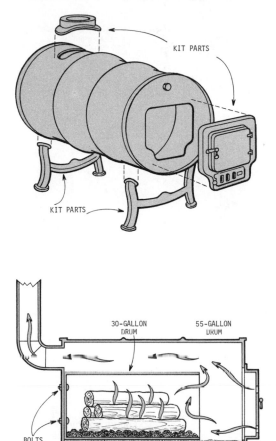

KIT PARTS

KIT PARTS

30-GALLON DRUM

55-GALLON DRUM

BOLTS

SAND FILLER

Kits for turning 55-gallon drums into large stoves are available from several sources. You can improve the performance of these stoves by slipping a 30-gallon drum inside to achieve a Scandinavian-stove draft pattern as shown here. It is also possible to install a thermostatically controlled intake by purchasing replacement parts from stove makers such as Martin Industries (Ashley).

can build an automatic drum heater, making use of a thermostat. Whether you try for a Scandinavian draft pattern or an Ashley-type thermostat-controlled stove, be sure to use furnace cement at all joints to assure air-tight construction.

HOW TO INSTALL A WOODSTOVE

Woodstoves give you more than lots of cheap heat. They give you easy installation. Once you've decided on a stove, and rounded up the parts and equipment you need to install it, the whole job shouldn't take you more than a day. Give yourself two days if you're not particularly good with tools. But don't let ease of installation fool you. The job should be done right. If you cut corners you'll be playing with fire. Chances are, the stove you buy will come with installation instructions. Most I've seen are fairly incomplete; some are not as conservative in the safety area as they could be. The advice that follows is based on the recommendations of the National Fire Protection Association (NFPA) and is safe and sound in all respects. NFPA recommendations are the basis of many fire and building codes throughout the country. If you follow them you should have no problems satisfying your building or fire inspector.

Before you begin any installation, check with your building inspector. Have your plan worked out in detail with a sketch of the installation, including clearances of stoves, stove

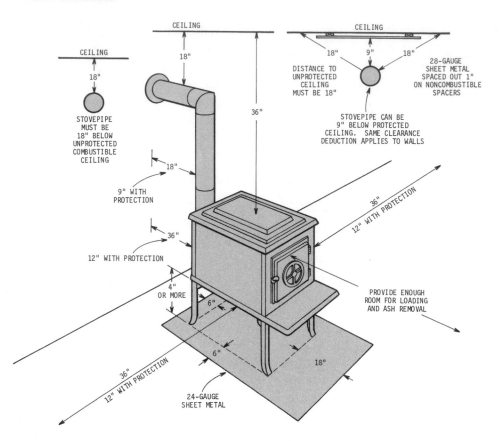

This shows minimum allowable clearances between radiant-type stoves and combustible surfaces. Combustible surfaces include plaster or wallboard construction, even though you might not think of these materials as combustible. Clearances shown include both protected and unprotected surfaces. Approved protection consists of 28-gauge sheet metal spaced one inch from the wall on noncombustible spacers. The 24-gauge sheet metal under the stove may be covered with masonry such as patio blocks or flagstones set in mortar if desired. This provides a more finished look. Remember that these clearances are minimums, and don't forget that furnishings such as furniture and curtains can be extremely combustible and can benefit from even greater clearances.

pipes, and chimneys from flammable walls, ceilings, furniture and floors. While dealings of this type can be a hassle in some cases, the hassle is less than the one you'll experience if the inspector makes you redo the work, or if the installation is unsafe and causes a fire. In some areas of the country you'll have no codes and no inspectors to satisfy. This spares you any encounter with the bureaucracy, but it also places responsibility for a safe installation solely in your hands.

Stove Location. First thing to consider in any stove installation is the loca-

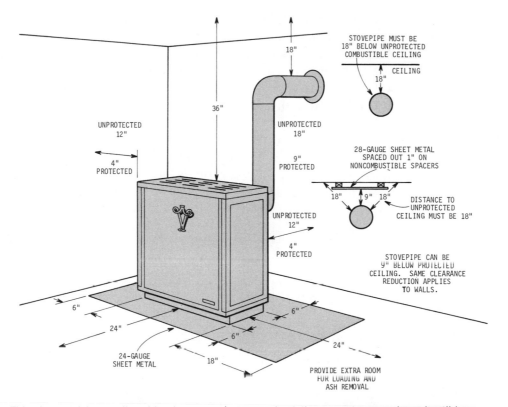

STOVEPIPE MUST BE
18" BELOW UNPROTECTED
COMBUSTIBLE CEILING

CEILING

18"

UNPROTECTED
12"

4"
PROTECTED

UNPROTECTED
18"

9"
PROTECTED

36"

18"

28-GAUGE SHEET METAL
SPACED OUT 1" ON
NONCOMBUSTIBLE SPACERS

18" 9" 18"

DISTANCE TO
UNPROTECTED
CEILING MUST BE 18"

UNPROTECTED
12"

4"
PROTECTED

STOVEPIPE CAN BE
9" BELOW PROTECTED
CEILING. SAME CLEARANCE
REDUCTION APPLIES
TO WALLS.

6"

24"

24-GAUGE
SHEET METAL

6"

18"

6"

24"

PROVIDE EXTRA ROOM
FOR LOADING AND
ASH REMOVAL

This shows minimum allowable clearances between circulating-type stoves and combustible surfaces. As noted on the facing page, combustible surfaces include plaster or wallboard construction.
Clearances are shown for both protected and unprotected surfaces. Approved protection consists
of 28-gauge sheet metal spaced one inch from the wall on noncombustible spacers. The 24-gauge
sheet metal under the stove may be covered with masonry such as patio blocks or flagstones set in
mortar. Note that many circulating stoves are loaded through the side. Remember to provide enough
room for loading and ash removal, and be sure to extend protective floor covering 18 inches on the
loading side of the stove.

tion of the stove. Where you put a stove will be determined by four factors:

1. Where will the stove do the best job of heating? In most cases this will be near the center of the area to be heated, and near the lowest point within that area. A central location makes for the most even distribution of heat, a low location helps take advantage of the fact that warm air rises. Keeping the stove on your first floor or even in the basement also makes for longer chimneys, better draft, better stove performance.

2. Where will the stove look best?

For the sake of aesthetics you may decide to locate your stove some distance from its ideal position for heating. A stove that totally dominates a room or impedes traffic is a problem no matter how well it heats. Remember, too, that not all parts of your home need the same amount of heat. Kitchens often provide most of their own. Bedroom and activity rooms can be cooler than rooms where you may sit and read, generating little body heat of your own.

3. Where will you vent the stove? All woodstoves require a chimney. If your home already has one, you may be able to hook into it. If so, you'll want your stove fairly close to the existing chimney. Or you may wish to vent your stove through an existing fireplace opening. Again, you'll have to put the stove close to the fireplace.

Converting a fireplace to woodstove operation is one good way to provide plenty of useful heat. The author's tiny stove, shown here, provides about two-thirds of the heat in his 1,200-square-foot home. The fireplace had been inefficient. The brick veneer wall behind the stove is a heat shield for the wall behind it. The stove stands on a 1¼-inch thick masonry hearth laid up with flagstones set in mortar. The copper-covered plywood woodbox has an upholstered top and holds a several-day supply of wood.

The author began with the conventional fireplace, at left. In order to shield the original wooden wall from heat, the author built a brick false front to a height of 5 feet. The first step was to mount sheet metal on the wall. Then with a 1-inch circulation space between the original and the new wall, a single column of bricks was built up each side. The columns were set 3 inches outside the original opening and were built to a height one brick higher, to allow using the old opening as a frame to later fasten a 1/8-inch sheet-metal plate and decorative copper flashing. A 3-inch angle-iron lintel was fastened across the tops of the columns to support the rest of the bricks. Part way up, brick angle ties were mounted. Remaining steps are evident in the drawing below.

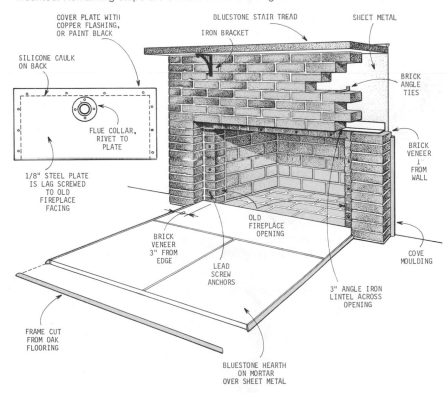

COVER PLATE WITH COPPER FLASHING, OR PAINT BLACK

BLUESTONE STAIR TREAD

SHEET METAL

IRON BRACKET

SILICONE CAULK ON BACK

BRICK ANGLE TIES

FLUE COLLAR, RIVET TO PLATE

1/8" STEEL PLATE IS LAG SCREWED TO OLD FIREPLACE FACING

BRICK VENEER 1" FROM WALL

OLD FIREPLACE OPENING

BRICK VENEER 3" FROM EDGE

LEAD SCREW ANCHORS

COVE MOULDING

3" ANGLE IRON LINTEL ACROSS OPENING

FRAME CUT FROM OAK FLOORING

BLUESTONE HEARTH ON MORTAR OVER SHEET METAL

Here are hardware and techniques used to install stoves in a variety of situations. Shown left to right are venting into an existing chimney; through a typical attic and roof; through an existing wall; and through a chalet-type ceiling.

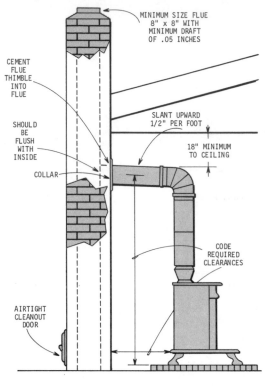

CEMENT FLUE THIMBLE INTO FLUE

MINIMUM SIZE FLUE 8" x 8" WITH MINIMUM DRAFT OF .05 INCHES

SHOULD BE FLUSH WITH INSIDE

SLANT UPWARD 1/2" PER FOOT

18" MINIMUM TO CEILING

COLLAR

CODE REQUIRED CLEARANCES

AIRTIGHT CLEANOUT DOOR

INTO EXISTING CHIMNEY

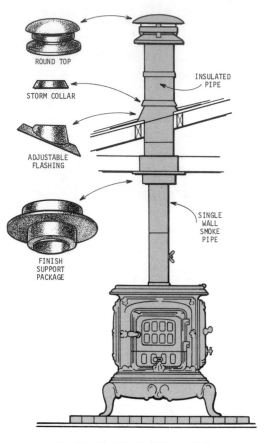

ROUND TOP

STORM COLLAR

INSULATED PIPE

ADJUSTABLE FLASHING

SINGLE WALL SMOKE PIPE

FINISH SUPPORT PACKAGE

THROUGH ATTIC AND ROOF

If you have no chimney at present, your best bet is to install one of the prefabricated metal types. These are very safe, and quite easy to install. You can run them up the outside of your home, or up through the ceiling (or through two or more stories) and on out the roof. If you go this route you'll have to think about where the chimney will run. Consider location of house framing members, walls, furniture in floors above the stove, trees, and overhead powerlines.

4. How can you meet minimum clearances? For safety, woodburning stoves must be installed certain minimum distances from any combustible surfaces. These minimum clearances—shown in drawings on pages 200 and 201—are the final factors you should consider.

The final location you choose will no doubt be a compromise that takes all four of the above factors into consideration. The object is to arrive at a safe, comfortable, eye-pleasing loca-

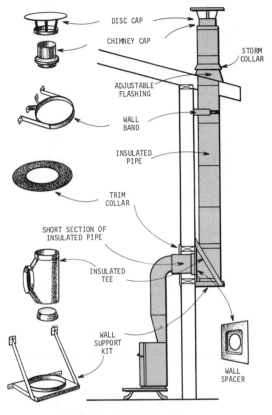

THROUGH EXISTING WALL

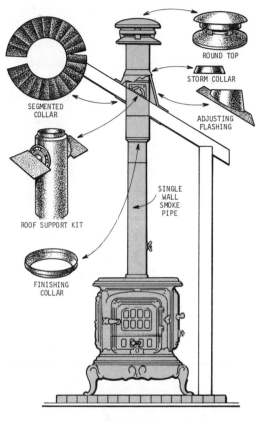

THROUGH CHALET-TYPE CEILING

tion, and one that won't create a lot of installation problems.

Installation from Scratch. To help you plan your stove installation, study the accompanying illustrations. These show how the most common hookups are made. One of these drawings may approximate your situation. Use it as a guide for planning your work. By making a similar sketch of your own with critical dimensions written in, you'll be able to show what you need

to your supplier, and he'll be better able to help you choose the right components for the job. Unless you're simply replacing an old stove with a new one you'll need a variety of supplies ranging from stovepipe to prefab chimney sections, and specialized chimney fittings.

WOODBURNING SAFETY TIPS

1. Start with good equipment, then inspect it yearly. Check stoves for

cracks, weak legs, faulty hinges.

2. Observe NFPA regulations. Allow for proper clearances between stove, stovepipe, and combustible surfaces. Remember that clearances shown in the installation sketches apply to wooden walls. Draperies and furniture are more easily ignited and require even greater clearances.

3. Always vent your stove to a proper chimney. This should be a masonry or brick chimney, lined with tile or mortar. Or a prefabricated Class "A," UL-approved flue. The chimney should be straight from top to bottom, and smooth inside — without ledges to collect ash and creosote. It should extend three feet above the point where it exits the roof, and two feet above any point on the roof within 10 feet.

4. Inspect your chimney before each heating season. Look for loose mortar, cracks, missing bricks, and soot and creosote. Repair cracks with concrete, and clean if necessary with a burlap bag stuffed with straw or paper and weighted with stones. The bag should fit the flue fairly snugly. Tie a rope to the bag, then lower it down the chimney and raise it back up several times. Cover the fireplace opening, or close the stove draft intake, to keep dirt out of your house.

5. Inspect the stovepipe for corrosion. Clean before every season, and be sure all joints are fastened with three sheet metal screws.

6. Check to be sure damper is operating properly and that it can't accidentally close. If it does so while you are asleep you could stay that way indefinitely.

7. Never pass a single-wall stovepipe through a wall or floor without the proper protection. Ventilated thimbles are good; multi-walled insulated stovepipe is even better. Stovepipe must rise at least $1/4''$ per foot of length. Avoid more than two sweeping 90-degree turns.

8. Avoid creosote production. Burn dry woods, and favor hardwoods over soft. Some stove makers recommend a chimney fire once a week to avoid buildups that could cause a larger chimney fire. To do this, burn a few sheets of crumpled newspaper with damper and draft controls wide open. Commercial salts, made for cleaning chimneys, can be thrown on the fire once a week or so to achieve the same results without requiring a chimney fire.

9. Don't overfire or underfire your stove. A roaring blaze creates too much heat and can ignite nearby walls or furniture. A smoldering fire creates creosote.

10. Don't burn trash or manmade logs.

11. Provide fresh air for the fire and the occupants of your home. Unless your home is particularly tight, however, you'll get enough air through cracks in the structure of the house.

12. Use kindling to start fires, not gasoline or other hydrocarbons.

13. Connect only one stove per flue. If you connect a stove to a flue above a fireplace, close off the fire-

place opening with a sheet of metal or asbestos boards. Make sure the sheet closes the opening snugly.

THE WOOD ITSELF

Just as there are good stoves and bad stoves, there are good and bad woods to burn in them. The best are the dense, heavy woods. The worst tend to be light. Good fuel woods burn slow, hot and clean, giving you more heat and less creosote per cord than the bad woods. A cord of hickory, for example, is equivalent to about two cords of white pine. That means hickory will take half as much cutting, handling, stacking, and hauling to heat your home as pine will. It will also take half the storage space. So it makes sense to choose fuel woods carefully, whether you cut your own or pay for it.

The accompanying table on fuel values of American woods lists our common fuel woods in approximate order of fuel value. At the top of each column is the price range you can pay for woods in that column and still heat your home for less than you could with 45-cent-a-gallon fuel oil. These same wood prices will also heat your home for about half the cost of 3.2-cent kilowatt-hour electricity. Woods at the top of each column give more heat than those at the bottom. So you can pay the prices at the high end of the range for woods near the top of the column. Or pay the low end of the price range for those near the bottom of the column. In either case, your

FUEL VALUES
OF VARIOUS AMERICAN WOODS

What Woods to Burn and How Much to Pay
Relative to Oil at 45¢/Gallon

$64-$77 PER CORD	$51-$64 PER CORD	$39-$51 PER CORD
High	Medium	Low
Live oaks	Holly	Black spruce
Shagbark hickory	Pond pine	Hemlocks
Black locust	Nut pines	Catalpa
Dogwood	Loblolly pine	Red alder
Slash pine	Tamarack	Tulip poplar
Hop Hornbeam	Shortleaf pine	Red fir
Persimmon	Western larch	Sitka spruce
Shadbush	Junipers	White spruce
Apple	Paper birch	Black willow
White Oak	Red maple	Large tooth aspen
Honey locust	Cherry	Butternut
Black birch	American elm	Ponderosa pine
Yew	Black gum	Noble fir
Blue beech	Sycamore	Redwood
Red oak	Gray birch	Quaking aspen
Rock elm	Douglas fir	Sugar pine
Sugar maple	Pitch pine	White pine
American beech	Sassafras	Balsam fir
Yellow birch	Magnolia	Black willow
Longleaf pine	Red cedar	Cottonwood
White ash	Norway pine	Basswood
Oregon ash	Bald cypress	Western red cedar
Black walnut	Chestnut	Balsam poplar

Prices at the top of each column are what you can pay for the woods in that column and still spend less than you would to heat your home on 45 cent a gallon oil. The high end of each price range pertains to woods at the top of each column. The low end pertains to those near the bottom.

heating costs will be about the same. Again, this is because you'll have to use more of the cheaper stuff to get the same amount of heat.

And remember: Some of the really bad woods, those down low in the right-hand column, are so hard to burn they may not do the job at all in fireplaces or bad stoves unless they are very dry.

Wood Facts. There are two ways to get wood. You can buy it or you can cut your own. Buying is relatively simple, so we'll cover that first. Gener-

GREEN/DRY WOOD HEAT VALUES

	Available heat per cord		Percent more heat from air-dry
	Green	Air-Dry	
	(Million BTU)		(%)
Ash	16.5	20.0	21
Aspen	10.3	12.5	25
Beech, American	17.3	21.8	26
Birch, yellow	17.3	21.3	23
Douglas-fir, heartwood	13.0	18.0	38
Elm, American	14.3	17.2	20
Hickory, shagbark	20.7	24.6	19
Maple, red	15.0	18.6	24
Maple, sugar	18.4	21.3	16
Oak, red	17.9	21.3	19
Oak, white	19.2	22.7	18
Pine, eastern white	12.1	13.3	10
Pine, southern yellow	14.2	20.5	44

The importance of burning dry wood is shown in this chart from the U.S. Forest Products Laboratory of the Forest Service. In addition to giving more heat, dry wood burns cleaner and produces less creosote. It's also easier to burn, and lighter and easier to handle. Wood will generally air dry in a year if kept covered. Split and bucked to short lengths it may dry in six months. Air circulation hastens drying.

EQUIVALENT TO
A CORD OF AIR-DRY WOOD

	Coal (tons)	#2 Fuel Oil (gallons)	Natural Gas (100 cu. ft.)
Ash	1.10	154	251
Aspen	.69	97	158
Beech, American	1.20	169	274
Birch, yellow	1.18	166	268
Douglas-fir, heartwood	.77	108	175
Elm, American	.93	130	211
Hickory, shagbark	1.36	191	309
Maple, red	1.02	144	233
Maple, sugar	1.18	166	268
Oak, red	1.18	166	268
Oak, white	1.26	176	286
Pine, eastern white	.67	94	152
Pine, southern yellow	.83	117	190

What is a cord of wood worth to you? Multiply the price you pay for coal, oil, or gas times the figures listed here to find out. Example: A cord of ash is worth about $70.30 if you pay 45 cents a gallon for oil. (154 gallons times 45 cents = $70.30) *Courtesy U.S. Forest Products Laboratory.*

ally, you buy wood by the *cord*. This is a stack of wood four feet high, four feet wide, and eight feet long. Theoretically this is 128 cubic feet of wood, but because of the spaces between logs, it works out to a lot less. Short, straight, thin logs will pack tighter than long, crooked, and thick logs. So will a mixture of different diameters. If you have a choice, buy wood that packs tighter and you'll get more wood per cord—maybe as much as 90 cubic feet out of that theoretical 128.

Not all cords are real cords. In many areas it's legal to call a *face cord* a cord. And a face cord is a stack of wood four feet high, eight feet long, but only as wide as the length of the logs. If the logs are two footers you'll only be getting half a true cord. So

know what you're getting before you buy.

When you're quoted a price for wood it can mean three different things. Undelivered price, delivered price, or delivered and stacked. Here again, know what you're getting. And in this same vein of caution, you should know if the wood is dry or green, and split or unsplit. You can usually recognize dry wood by looking at it. The ends of the logs should not look freshly cut. Cracks or checks should be visible on the log ends. A couple of logs smacked together should ring rather than thud dully. If you drive an axe or wedge into the end of a log it shouldn't ooze water. If you have the time to let green wood season (at least a few months and a year is better) go ahead and buy it green. But if you want to burn it immediately, buy seasoned wood. Green wood gives less heat and more creosote.

Now, about splitting. Big logs must be split if they are to burn well. Some will have to be split just to fit into your stove. Figure that anything over five inches should be split. Any log over eight inches should be split into quarters, and any log over a foot will take a lot of splitting. If you enjoy that kind of work, buy unsplit wood. Otherwise, split wood will save you work. And since most suppliers do their splitting with machines, splitting usually won't add much to the costs.

Cutting Your Own.   The first trick to cutting your own wood is finding it. If you have your own woodlot, fine. If not there are still plenty of places to get wood. One quick place to check is your local landfill or dump. Many tree surgeons and town crews take the wood they cut on the job right to the dump. Road-building projects often create a lot of free firewood, both in the form of trees cut to clear the right of way and in waste lumber from forms, temporary rigging and the like. Sometimes you can even find people who'll pay you to cut and take away dead or unwanted trees on their property.

Recently, the U.S. Forest Service and the Bureau of Land Management adopted a policy that allows free cutting of firewood on federal lands. Before you can cut you must apply for a permit at any National Forest district office. This gives the Service a chance to show you where and what to cut. There is a maximum limit of ten cords a season, and you can't sell the wood; you have to burn it yourself. But if you need more than ten cords to heat your home, something is wrong somewhere.

Much of the wood you're allowed to cut will be in the form of logging

CHARACTERISTICS OF VARIOUS FIREWOODS

HARDWOOD TREES	Relative amount of heat	Easy to burn	Easy to split	Does it have heavy smoke?	Does it pop or throw sparks?	General rating and remarks
Ash, red oak, white oak, beech, birch, hickory, hard maple, pecan, dogwood	High	Yes	Yes	No	No	Excellent.
Soft maple, cherry, walnut	Medium	Yes	Yes	No	No	Good.
Elm, sycamore, gum	Medium	Medium	No	Medium	No	Fair—contains too much water when green.
Aspen, basswood, cottonwood, yellow-poplar	Low	Yes	Yes	Medium	No	Fair—but good for kindling.
SOFTWOOD TREES						
Southern yellow pine, Douglas-fir	High	Yes	Yes	Yes	No	Good but smoky.
Cypress, redwood	Medium	Medium	Yes	Medium	No	Fair.
White-cedar, western redcedar, eastern redcedar	Medium	Yes	Yes	Medium	Yes	Good—excellent for kindling.
Eastern white pine, western white pine, sugar pine, ponderosa pine, true firs.	Low	Medium	Yes	Medium	No	Fair—good kindling.
Tamarack, larch	Medium	Yes	Yes	Medium	Yes	Fair.
Spruce	Low	Yes	Yes	Medium	Yes	Poor—but good for kindling.

Courtesy U.S. Forest Products Laboratory

wastes. You'll want to pick over this stuff pretty carefully, looking for the logs that will be easiest to cut and split. Avoid logs covered with mud and gravel, for they'll dull your saw almost immediately. Many of the logs will be under tension, twisted and intertwined with other logs. Caution is in order or you could release the tension and be seriously injured by the resulting kick of the log. More about this later on.

Wherever you cut your wood, try to cut the species with the highest fuel values. If you don't know how to identify trees, a good field guide will help. One popular guide for use when trees are in leaf is *Trees of North America*, published by Golden Press. But for exacting identification, you may need a more thorough book, such as *A Field Guide to Trees and Shrubs* by Houghton Mifflin Company; this provides both summer and winter identification keys.

Woodcutting Tools. The basic tool for cutting wood is the chain saw. You can use an axe or a bow saw for cutting small amounts of wood, but anything over a cord or so a year justifies a chain saw. A medium-size saw with a 14-inch bar should do the job in almost all cases, though you might want to go to a 20-inch bar on a larger saw if you have a lot of big trees to cut. I use a lightweight saw with a 14-inch bar and it works like a charm. The secret is a sharp chain. When you buy a saw, get a chain file at the same time, and use it. A sharp chain will eagerly eat

its way into a log. You'll have to force a dull saw to get anywhere at all.

Any gasoline chain saw is loud, large or small, so once I fell a tree I like to remove all but the largest branches with a light axe. This not only provides relief from the noise of the saw, but single-stroking off branches with a sharp axe is as fun as driving a golf ball. Again, sharpness is the key. Use a file or grinder to cut a sharp edge and remove any nicks. Then go over the blade with an axe hone. A round one with coarse and smooth sides is a handy type to keep in your pocket. Once you get the axe sharp, keep it that way. It's easier than resharpening a blade that has lost its edge. Remember, a sharp axe is easier to work with, and safer than a dull one. It's the dull axe that skids off the work and bites into your leg.

This axe is not the same one you'll use later on for splitting. Use it only to knock off branches and fell an occasional tree. Splitting is better done with an axelike tool called a splitting maul or hammer. It looks like a sledge hammer with a wedge on one end of the head, a flat hammer face on the other. The fat wedge edge splits wood better than an ordinary axe, and is less likely to stick in the end of a log. The splitting maul plus a couple steel wedges make up your full splitting team. You need that second wedge in case the first gets stuck—a pretty common occurrence if you bother with knotty, twisted logs. These wedges should be steel. Aluminum, wood, and plastic wedges you see in stores should be used only for freeing stuck chain saws. You'll probably find one of these handy when you're bucking logs to firewood length, but don't use them for splitting. Their soft makeup—meant to protect the chain saw's teeth—can't take the punishment of splitting.

Tool Use. First step, of course is felling. Almost all trees prefer to fall in a specific direction. You can make them fall some other way at times, but not often. Best bet is to go along with the wishes of the tree. Look it over and see which way it's leaning. If it's standing straight, see which side has the heaviest load of branches. That's the way the tree will want to go. Make a wedge cut into that side of the tree, then a horizontal cut from the other side, slightly above the apex of the wedge, and the tree should fall by itself.

Start the wedge cut with a horizontal cut about a third of the way through the tree. Then make a second cut down towards the first to cut out the wedge. With practice you'll be able to cut the wedge with two quick cuts. But at first it may take a little fiddling to make the two cuts intersect. Once the wedge is out, start the felling cut from the opposite side of the tree. Make the cut horizontal, aiming for the spot an inch or two above the horizontal wedge cut. As the felling cut approaches the wedge, watch for signs that the tree is starting to fall. As soon as it makes its move,

you make yours. Pull the saw from the kerf and back off. A falling tree can jump at you off the stump, and you don't know what a tree will do once it hits the ground. It may kick back in its death throes, taking you with it.

If the tree is in leaf, you may want to leave it for a few weeks, until the leaves dry. The tree will continue to transpire, or pass water out through its leaves, and this speeds drying. Once the leaves turn brown they're through passing off water and you can begin bucking. First step here is to knock off the small branches with your axe. Start at the base of the tree and work towards the top. Your axe cuts should be in this same direction. Cut the branches as close to the trunk as possible. Any stubs will interfere with stacking and efficient stove loading later on. Larger branches are easier to remove with a chain saw. Before you knock off these bigger branches, check to see how the tree will react. These branches are supporting the bulk of the tree's weight, and when you remove them the tree may roll or jump your way. Large branches make good fuel, so strip them of smaller branches with your axe, then treat them just like the main log.

Now comes bucking, and the time for a little planning. You want to plan your work from now on to avoid a lot of wasted effort. If you can drive a pickup truck to where the log lies waiting, it's pretty easy to buck the log into firewood lengths and toss

The first step in felling a tree is to take a wedge-shaped chunk of wood out of the tree. Make sure this chunk is cut from the side of the tree facing the direction you want the tree to fall. For a tree with a vertical trunk and good symmetry, the wedge cut should go about one-third of the way through the trunk.

The felling cut is horizontal and from the side opposite the wedge cut. Aim the felling cut an inch or two above the wedge cut. When the tree starts to fall, remove the saw and back off, for the tree may kick back towards you while falling or after it hits the ground.

these into the truck. Really thick logs will be too heavy to move and will require splitting on the spot. The big hassle in bucking a downed tree is the tree's insistence upon pinching the chain saw bar in the kerf. You can avoid this for the most part by raising the trunk slightly near the intended cut by means of large wooden wedges or a piece of timber. Then cut the log to lengths you can handle before using a sawbuck for bucking to final length. Always cut down through the log as it hangs over the end of the

sawbuck, and you'll never get the saw stuck. Small tip: Cutting through big knots and burls makes splitting easier later; this is especially important for tougher splitting woods.

I prefer not to buck up my firewood until I have it where I want it. My typical log is usually around six inches in diameter, and I find it simpler to handle this size stuff in lengths around six or eight feet. You may not find it easier to handle a single eight-footer as opposed to four two-footers, but I do. However you

Next lop off the limbs. Some people wait a few weeks before removing branches. This gives the leaves a chance to draw moisture from the tree before they finally wither, thus drying the wood further.

Firewood is usually easier to carry out in lengths than it is when cut to stove length. But large trees can't be handled this way.

A sawhorse like this light-duty one speeds the bucking operation. Work goes even faster if you have a helper to place logs in cutting position.

Splitting tools include an axelike splitting maul, wedges, and a sledge hammer. Splitting is easiest when you work on a large chopping block about two feet high. The block brings your work up to a convenient height, provides a solid base, and prevents your maul's cutting edge from hitting stones and dirt.

Easy-splitting wood can usually be handled with a single stroke of the maul. Be sure you have a good footing, with feet wide apart for best control.

handle your wood, try to avoid repeating steps. A good sequence would be: Fell, debranch, cut to convenient lengths, haul to final storage site, buck, split, stack. Stacking at the cutting site is a waste of time. So is loading and unloading a lot of short logs.

If you have a lot of bucking to do, it may be worth while to make a bucking crib. This is a double row of fence posts driven into the ground. You space the two rows about one log diameter apart. Stack several logs in be-

tween the rows, then use your chain saw to cut the logs to length by slicing down through the stack. This is a lot faster than any other bucking method, and if your soil will let you drive the posts without a hassle you can make a crib almost as fast as a sawbuck.

Splitting will speed seasoning. So if you're in a hurry to put new wood to use, split before stacking. Some woods are harder than others to split (table on page 209), but none is impossible. Unfortunately for the wood splitter, some of the most readily

If the maul sticks in the log, you can drive it on through like an ordinary wedge, using a sledge hammer. The handle of the sledge here has been fiberglassed and wrapped with twine while the resin was still wet. This reinforces the handle at its weakest point.

bers won't, so it's best to attack these with the wedge. If the maul sticks in the log, freeing it is a lot of work and a waste of time. So use the maul to drive the wedge—or wedges if the first wedge sticks out of sight in the log. Whether you're driving wedges or attacking directly with the cutting edge of the maul, let the weight of the tool do the work. Raise the tool and let it drop, keeping your eye on the target. A little muscle added to the weight of the maul won't hurt. But as soon as you start trying to overpower the operation you're inviting trouble in the form of broken hammer handles or worse.

If your logs have started to dry, they'll have natural check lines in their ends. Use these as guides for your splitting. Sometimes they'll be wide enough to hold your wedge, other times you'll have to start the wedge as you'd start to drive a nail. Sometimes wood splits easier if it's very cold. But I've found that very wet wood that freezes causes problems. The frozen water in the wood lubricates the wedge. You drive it in, it squirts back out like a watermelon seed.

If you have a lot of wood to split, consider renting a mechanical splitter. Or if you have a tremendous amount of splitting (for a commerical operation), consider buying.

available woods are hardest to split. There are a lot of elms—victims of Dutch elm disease—good for little else than burning. But they're the devil to split. And the easiest logs to split are usually the clear, straight-grained parts of trees that can be better used for lumber.

Despite all this, splitting can be fun if you take it slow and easy. Don't try to do too much at one time, don't rush, and don't force the tools. Easy logs will succumb to a single stroke from your maul. The tougher num-

Storing Wood.   Stacking is more than an orderly way to store wood. It also speeds drying, and protects against rot. For fastest drying, stack your wood in a grid pattern. Place two or three sticks parallel to one another, then two or three on top of these at right angles, and so on in a chimney-like stack. This allows for better air circulation between sticks. And wood stacked this way is less likely to roll out of the stack. Keep rain and snow off the stack by putting it under cover: either a shed, a simple sheet metal roof, or plastic film. Raising the stack off the ground on wood or concrete blocks will also speed drying and prevent the bottom of the stack from rotting.

Wood cut in the spring, then split and stacked, will be acceptable for burning by the time winter comes. An extra year of drying in the barn, basement, or shed is even better. But few people are disciplined enough to work that far in advance of their needs. Anyway, spring is a nice time to work. The weather isn't too hot, the woods aren't overgrown with weeds, the bugs are yet to arrive in force. Flowing sap does, however, tend to make the wood swell as you saw, closing the kerf and pinching your saw.

The most convenient time to cut wood, however, is during the winter. The snow on the ground simplifies transportation, especially if you can cut wood uphill from where you'll use it. A sled or toboggan loaded with wood will slide downhill all the way to your woodpile. About all you have to do is guide it in the right direction and keep it from speeding out of con-trol. Even if you have to drag the wood on the level or uphill, the sled or toboggan is still the easiest way to go. Both are low to the ground, so loading is easy. With the toboggan you can just roll logs into place without any lifting at all. With a good cover of snow on the ground, terrain that's too rough for wheeled vehicles in summer often levels out nicely for easy sliding. Early Americans knew well the advantages of moving heavy loads during the winter. In fact they saved that kind of work for the snow season. Heavy rocks, big stumps, firewood, sawlogs were all moved by sliding.

Your Own Woodlot. According to the U.S. Forest Service, the average acre of forest can produce about a cord of wood a year. So a woodlot equal in acres to the number of cords you burn in a year will keep you in firewood for life. You can figure on five or so cords as an average yearly need—more for big drafty homes in cold climates, less for small, well-insulated homes. Of course some of the wood you grow may be more valuable as lumber than firewood. If you plan to use it as such your cords-per-acre yield for firewood will drop, and you'll require a larger woodlot.

Yields higher than a cord per acre per year are possible with special techniques. Sprout forests are one example. Trees are allowed to grow for a number of years—say 12—then cut for wood. Their growth during these first 12 years is fast since the trees are young and vigorous. After they are cut, sprouts pop out of their stumps. You prune to encourage a single sprout which then grows rapidly, living a rich life off the roots established by the first tree. The advantages are rapid growth and small logs that require no splitting.

Sawlogs or Firewood? You'd be an utter fool to burn everything that grows on your woodlot. Good lumber is always worth more than firewood. So it pays to learn how to recognize valuable species and how to tell a sawlog from a chunk of stove fodder.

The most valuable trees are the quality hardwoods—such as birch, oaks, sugar maple (not red maple), cherry, basswood, ash, red gum, and walnut. Walnut has become so valuable that individual trees are being stolen by fast operators with chainsaws and trucks. The money you make from a single walnut sold for lumber will buy you many more Btu worth of firewood than you'd ever get from the walnut.

Good sawlogs usually run at least eight feet long, although the more valuable the wood, the shorter the log can be. Minimum diameter should be about eight inches. While straight logs are usually most valuable, don't forget that crotches, butts and burls are the source of highly-figured woods used in premium veneers.

SOURCES FOR WOOD HEATING EQUIPMENT

Stoves, Furnaces, Ranges

Ashley Automatic Heater Company
1604 17th Avenue, S.W., Box 730
Sheffield, AL 35660

Ashley space heaters

Atlanta Stove Works
Atlanta, GA

Cast-iron stoves

Autocrat Corporation
New Athens, IL 62264

Ranges and space heaters

Bell, Perley
Grafton, VT 05146

Stoves and furnaces

Birmingham Stove and Range Company
Box 3593
Birmingham, AL 35202

Stoves, heaters, and ranges

Brown Stove Works, Inc. Cleveland, TN 37311	Space heaters
C & D Distributors, Inc. Box 766 Old Saybrook, CT 06475	Steel box stove for fireplace installation
Cummings, Waldo G. Fall Road East Lebanon, ME 04027	Furnaces
The Dam Site Stove Company R.D. No. 3 Montpelier, VT 05602	The Dynamite stove
Sam Daniels Company Box 868 Montpelier, VT 05602	Wood burning furnaces
Duo-Matic 2413 Bond Street Park Forest South, IL 60466	Multi-fuel furnaces
Fawcett Division, Enheat Ltd. Sackville, New Brunswick, Canada	Cast-iron box, Franklin, parlor, and cookstoves
Fireview Distributors Box 370 Rogue River, OR 97537	Barrel stoves
Fisher Stoves 504 S. Main Concord, NH 03301	Fisher stoves
Garden Way Research Department 64456 Charlotte, VT 05445	Steel box stove
Gay, Larry Marlboro, VT 05344	Steel box stove, accessories
H. D. I. Importers Schoolhouse Farm Etna, NH 03750	Scandia range, Lange stoves

Heatilator Fireplace Division, Vega Industries, Inc. Mt. Pleasant, IA 52641	Heatilator fireplaces
Kickapoo Stove Works, Ltd. Main Street, Box 14W La Farge, WI 54639	Wood burning furnaces, stoves
King Stove and Range Company Box 730 Sheffield, AL 35660	All kinds of stoves
Kristia Associates Box 1118 Portland, ME 04104	Jotul line of stoves, ranges, fireplaces
Marathon Heater Company, Inc. Box 165, R.D. 2 Marathon, NY 13803	The Logwood Furnace
Mohawk Industries, Inc. 121 Howland Avenue Adams, MA 01220	The Temp-Wood stove
"Old Country" Appliances Box 330 Vacaville, CA 95688	Tiroha European range
Preston Distributing Company Whidden Street Lowell, MA 01852	Chappee stoves
Portland Franklin Stove Foundry Company Box 1156 Portland, ME 04100	Cast-iron stoves and ranges Trolla stoves
Ram Forge Brooks, ME 04921	Heavy steel box stove
Riteway Manufacturing Company Box 6 Harrisonburg, VA 22801	Riteway stoves, multi-fuel furnaces and boilers

Self Sufficiency Products
One Appletree Square
Minneapolis, MN 55420

Sierra, Alaskan and Gibralter IV
stoves

Shenandoah Manufacturing
Company, Inc.
Box 839
Harrisonburg, VA 22801

Shenandoah space heaters

Scandinavian Stoves, Inc.
Box 72
Alstead, NH 03602

Lange stoves, Tiba ranges

Sunshine Stove Works
Callicoon, NY 12723

Steel box stoves

Tekton Design Corporation
Conway, MA 01341

Kedelfabrik-Tarm imported wood-oil
boilers, Lange stoves

The Merry Music Box
10 McKown Street
Boothbay Harbor, ME 04538

Styria heaters and ranges

United States Stove Company
Box 151
South Pittsburgh, TN 37380

All kinds of stoves

Vermont Castings, Inc.
Box 126, Prince Street
Randolph, VT 05060

The Defiant stove

Vermont Counterflow Wood Furnace
Plainfield, VT 05667

Wood burning furnaces

Vermont Woodstove Co.
Bennington, VT 05201

DownDrafter stoves

Washington Stove Works
Box 687
Everett, WA 98201

Cast-iron stoves, and a variety of
cooking ranges, including marine
ranges

Wilson Industries
2296 Wycliff
St. Paul, MN 55114

Multi-fuel furnaces

Accessories

Blazing Showers
Box 327
Pt. Arena, CA 95468

Hot water heat exchangers

Calcinator Corporation
Bay City, MI 48706

Magic Heat stovepipe heat exchangers

Patented Manufacturing Company
Bedford Road
Lincoln, ME 01773

Slip-on stovepipe heat fins

Sturges Heat Recovery, Inc.
Stone Ridge, NY 12484

Thriftchanger stovepipe heat exchangers

Portland Willamette Company
6804 N.E. 59th Place
Portland, OR 97218

Glassfyre screens

Taos Equipment Manufacturers, Inc.
Box a565
Taos, NM 87571

Good $200, power log splitter

WATER POWER

Nearly all our large-scale water power sites have been exploited to their limits. But there are still plenty of ways to put water power to use on a small scale.

Water power is clean. If you're lucky enough to have a year-round stream on your property, water power is dependable. It's also constant, so there's no need to store your power as there is with solar and wind energy. Over the long haul hydroelectric power you generate yourself can even be cheap. Hydroelectric generators are almost maintenance free, so once your operation is running, your costs are just about zero.

But don't think for a minute that you can set up a water power unit without a large original investment.

The smallest hydroelectric turbine you can buy will cost over $4,000. Installed, the cost will rise to about $10,000. And all this for a unit that will produce only 500 watts of DC power. This brings up a point: Larger turbines produce much cheaper power than small ones. A 2,000 watt hydroelectric plant is only a few hundred dollars more than the 500-watter for $4,000. And a 10,000 watt model will run only around $7,500.

Obviously then, it doesn't pay to think too small. The only drawback with larger units is that they require more water power to run them. So picking the right unit is a matter of balancing your electrical needs against the water power available on your property. Here's how to start:

DETERMINING AVAILABLE POWER

The amount of power available in any stream is simply a combination of two factors. These are the volume of water flowing in that stream, and the "head" or distance the water falls. Head and volume interrelate; you can get the same amount of power from a little bit of water falling a long distance as you can from a lot of water falling a short distance. Example: A volume of 760 cubic feet per minute falling through 8 feet gives 5 kilowatts. So will a volume of just 260 cubic feet falling 25 feet.

How to Measure Flow. To do this right, you should determine flow both in early spring and late summer. This will provide flow rates when the stream is at maximum and minimum flow. There are two ways to measure flow. The simplest is the float method.

Pick a spot on the stream where the water flows smoothly. Measure the average depth and width of the stream. Then place a bottle partially filled with water in the stream. The bottle should be about 60 percent submerged. Let the bottle float downstream and measure the distance it travels in a minute. Now simply multiply the average width by the average depth, by the distance travelled, by the constant .8. Your result will be the approximate flow in cubic feet per minute.

As you can see, to be accurate with this system, you'll have to come up with the average depth and width of the stream over the entire distance the bottle travels during its one-minute run. This will be easiest where the stream is of a constant depth and width. If you can't come up with a spot like this, you'll have to make several measurements of width and depth throughout the length of the bottle's run. Then you'll have to average the measurements to come up with reasonably accurate width and depth readings.

Your final results may not be right on the nose, but in many cases they'll tell you whether or not your stream can be used for power generation. You can compare your flow figures against the rating table of Leffel's hydroelectric units on pages 226 and 227 to see if the stream fits your needs.

If your stream is a borderline case you can make a more accurate measurement by another method. Again you should start by finding a spot where your stream flows smoothly. Construct a weir of boards across the stream as shown in the drawing on the next page. For the best readings, the notch in the top of the weir should be about six times as wide as the depth of the water flowing over the weir. The edges of the notch should be beveled as shown, and the bottom of the notch should be at least six inches above the level of the water on the downstream side of the weir.

Next, drive a stake into the stream bottom about three feet upstream from the weir. The top of this stake should be dead level with the bottom of the notch. Now measure the depth

HOW TO MEASURE WATER FLOW

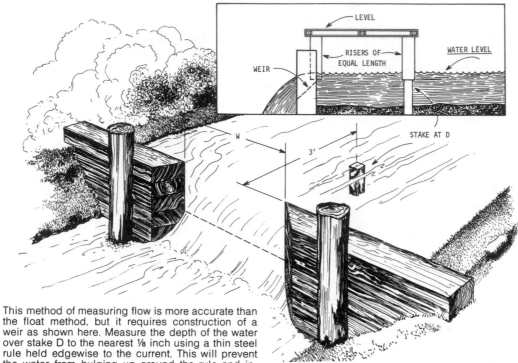

This method of measuring flow is more accurate than the float method, but it requires construction of a weir as shown here. Measure the depth of the water over stake D to the nearest ⅛ inch using a thin steel rule held edgewise to the current. This will prevent the water from bulging up around the rule and increasing the reading. The top of stake D must be dead level with the spillway. To check for this condition without dunking your level in the water, place wooden risers of equal length atop stake and spillway. Place your level atop these risers as shown in the inset. Find the flow rating in the table by first reading down the left column to find water depth in inches, then reading across the top row to find fractions of inches. Example: If the water is 10⅝ inches deep, read across the 10-inch row, down the ⅝ inch column. The result is 13.93. Multiply this figure by the width of the spillway opening (W) to find flow in cubic feet per minute.

INCHES DEPTH OVER STAKE D	WHOLE INCHES	⅛ Inch	¼ Inch	⅜ Inch	½ Inch	⅝ Inch	¾ Inch	⅞ Inch
1 Inch	.40	.47	.55	.65	.74	.83	.93	1.03
2 "	1.14	1.24	1.36	1.47	1.59	1.71	1.83	1.96
3 "	2.09	2.23	2.36	2.50	2.63	2.78	2.92	3.07
4 "	3.22	3.37	3.52	3.68	3.83	3.99	4.16	4.32
5 "	4.50	4.67	4.84	5.01	5.18	5.36	5.54	5.72
6 "	5.90	6.09	6.28	6.47	6.65	6.85	7.05	7.25
7 "	7.44	7.64	7.84	8.05	8.25	8.45	8.66	8.86
8 "	9.10	9.31	9.52	9.74	9.96	10.18	10.40	10.62
9 "	10.86	11.08	11.31	11.54	11.77	12.00	12.23	12.47
10 "	12.71	12.95	13.19	13.43	13.67	13.93	14.16	14.42
11 "	14.67	14.92	15.18	15.43	15.67	15.96	16.20	16.46
12 "	16.73	16.99	17.26	17.52	17.78	18.05	18.32	18.58
13 "	18.87	19.14	19.42	19.69	19.97	20.24	20.52	20.80
14 "	21.09	21.37	21.65	21.94	22.22	22.51	22.70	23.08
15 "	23.38	23.67	23.97	24.26	24.56	24.86	25.16	25.46
16 "	25.76	26.06	26.36	26.66	26.97	27.27	27.58	27.89
17 "	28.20	28.51	28.82	29.14	29.45	29.76	30.08	30.39
18 "	30.70	31.02	31.34	31.66	31.98	32.31	32.63	32.96
19 "	33.29	33.61	33.94	34.27	34.60	34.94	35.27	35.60
20 "	35.94	36.27	36.60	36.94	37.28	37.62	37.96	38.31
21 "	38.65	39.00	39.34	39.69	40.04	40.39	40.73	41.09
22 "	41.43	41.78	42.13	42.49	42.84	43.20	43.56	43.92
23 "	44.28	44.64	45.00	45.38	45.71	46.08	46.43	46.81
24 "	47.18	47.55	47.91	48.28	48.65	49.02	49.39	49.76

of the water over the top of the stake. Try to measure to the nearest eighth inch. Take the result and look it up in the weir flow chart, as explained in the caption accompanying the drawings.

Okay, now you have an accurate flow rate for your stream. And as mentioned, you should do this in the spring when flow is at its greatest, and in late summer when flow is low. If you want year-round power, the minimum flow rate is the one that matters. It doesn't matter much how good the flow is in the springtime if the stream slows to a trickle later in the year. Of course, most water power installations will have a dam and this can help you hedge somewhat on your low-flow calculations. This is because the dam can store up water when demands on the turbine are low, then release the water when power is needed. This storage can be short term (overnight) or long term (over a period of weeks) if the area of the water impounded is great enough.

Once you've determined the flow of your stream, you'll have to think about head. Normally, the more the better. If you are lucky you'll have a waterfall or at least a rapid drop in stream level that will provide you with all the head you need. If your stream flows gently over a long flat stretch, however, you'll have to create your own head by damming the water. This raises the water level. You can then locate your mill some distance downstream from the dam,

and feed the mill via a millrace. Naturally, the farther downstream you place the mill, the greater the head will be. But a long millrace creates friction that retards water flow. That's why you should try to locate your installation where there is a waterfall or fast drop in the stream bed. The closer your mill is to the dam, the better, and the cheaper the installation will be to build.

If you have a surveyor's transit you'll have no trouble using it to determine your head. If not, you can get good results by making a series of stepped measurements and then adding them together. The accompanying drawing shows how this is done. If you plan to build a dam, but haven't done so yet, at least try to

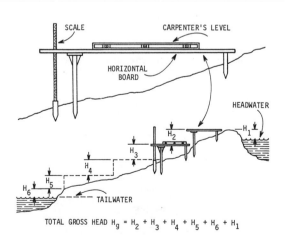

TOTAL GROSS HEAD $H_g = H_2 + H_3 + H_4 + H_5 + H_6 + H_1$

To measure head without a transit, use a carpenter's level as shown. Take a series of measurements and add them all together to get total head. If you have no dam yet, you won't have the true height of the headwater. Estimate the intended height of the dam, and consider the top of the dam as the elevation of the headwater.

decide where it will go and how high it will be. This will give you an approximate figure for the height of your headwater.

If you haven't decided where to place the mill, but do have an idea of how much head you want, add your stepped measurements together as you progress down the stream. When your measurements add up to the desired head, you know where you will put the mill.

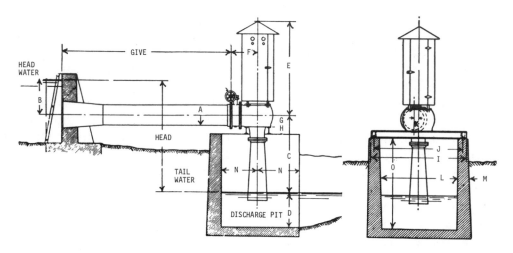

Ratings of the various Hoppes hydroelectric units, their dimensions, and requirements in head and flow are given here to help you with your preliminary planning. Other hydroelectric units are available from European sources, but Leffel is a convenient domestic source and their Hoppes units are well designed. Generators are direct-coupled to the turbine without the need for belts and pulleys or gearing. Working parts are enclosed for weather protection and are placed above the water level to prevent damage during floods. Costs will vary according to size, and the cheapest unit for a given rating is the one designed for the greatest head. This is because a greater head lets a smaller power plant accomplish the job. For more information, write James Leffel & Company, 426 East St., Springfield, OH 45501.

GENERAL DIMENSIONS

Style Sizes in Inches

	F	HJ	HL IL	JP	LR	OT
A	6	10	12	16	18	20
B	18	24	24	36	36	36
C	60	60	60	60	60	60
D	24	30	30	36	42	48
E	58	72	72	88	90	90
F	20½	18	18	22	23	27
G	5½	8⅝	9⅝	11⅛	11⅞	14¼
H	5	5	5	6	6	7
I	72	72	72	84	84	96
J	66	66	66	78	78	90
K	3⅛	4¾	4	4	4	4½
L	60	60	60	72	72	84
M	8	8	8	9	10	10
N	30	30	30	36	42	48
O	73½	73⅜	75⅜	78⅞	84⅛	86¾

Once you've determined head and flow, you can sit down and see how much power you can generate. Take a look at the rating table of Hoppes units and see where your stream fits in. See how many watts you can pro-duce. Now balance this against your desire for power. How many watts do you want? To do this, simply add up the wattage ratings of all the elec-trical devices you might conceivably run all at one time. To provide the

RATING TABLES OF HOPPES HYDRO-ELECTRIC UNITS

ELECTRICAL CAPACITY	Head in Feet	Style	Water in Cubic Feet Per Minute
½ KILOWATT OR 500 WATTS	* 8	HL	104
	* 9	HJ	92
	* 10	HJ	82
	* 11	F	74
	* 12	F	68
1 KILOWATT OR 1000 WATTS	* 8	IL	190
	* 9	HL	175
	* 10	HL	155
	11	HL	140
	12	HJ	127
	13	HJ	118
	14	HJ	110
	15	HJ	105
	16	HJ	100
	17	HJ	94
	18	HJ	90
	19	F	84
	20	F	80
	21	F	76
	22	F	74
	23	F	72
	24	F	70
	25	F	68
2 KILOWATTS OR 2000 WATTS	* 8	JP	330
	* 9	JP	290
	* 10	JP	260
	* 11	HL	245
	* 12	HL	225
	* 13	HL	215
	14	HL	190
	15	HL	178
	16	HL	166
	17	HL	156
	18	HJ	153
	19	HJ	148
	20	HJ	140
	21	HJ	133
	22	HJ	127
	23	HJ	120
	24	F	116
	25	F	110
3 KILOWATTS OR 3000 WATTS	8	LR	470
	9	JP	415
	10	JP	370
	11	JP	340
	12	JP	310
	13	JP	280
	14	JP	260
	15	IL	250
	16	IL	240
	17	HL	225
	18	HL	210
	19	HL	200
	20	HL	190
	21	HL	180
	22	HL	170
	23	HL	165
	24	HJ	162
	25	HJ	158

ELECTRICAL CAPACITY	Head in Feet	Style	Water in Cubic Feet Per Minute
5 KILOWATTS OR 5000 WATTS	8	OT	760
	9	OT	600
	10	LR	590
	11	LR	535
	12	LR	490
	13	JP	470
	14	JP	435
	15	JP	400
	16	JP	365
	17	JP	340
	18	JP	330
	19	JP	320
	20	JP	315
	21	HL	300
	22	HL	290
	23	HL	285
	24	HL	275
	25	HL	260
7½ KILOWATTS OR 7500 WATTS	11	OT	800
	12	OT	740
	13	LR	680
	14	LR	630
	15	LR	590
	16	JP	550
	17	JP	515
	18	JP	490
	19	JP	480
	20	JP	450
	21	JP	430
	22	JP	410
	23	JP	400
	24	JP	390
	25	JP	380
10 KILOWATTS OR 10,000 WATTS	12	OT	980
	13	OT	900
	14	OT	840
	15	OT	780
	16	LR	715
	17	LR	670
	18	LR	650
	19	JP	610
	20	JP	580
	21	JP	550
	22	JP	525
	23	JP	500
	24	JP	490
	25	JP	480

Direct-current units

kind of electrical service most Americans are used to, you might consider 3,000 watts as a minimum. This could run a furnace, refrigerator, a few lights, and a washer, with some reserve left over to handle the high loads caused whenever an electric motor starts up.

At this point you may have to reevaluate your situation. It's possible to change your lifestyle and cut electrical needs considerably. It helps to run power-hungry devices only when other demands are low. You can also help by starting motor-driven devices one at a time, starting with the largest. But some appliances such as refrigerators and freezers and furnaces are automatic and run according to orders from thermostats. There's only so much room for compromise.

AC/DC Options. You'll notice when you look over the Hoppes ratings table on the preceding pages that you can take your pick of AC or DC generators in some of the smaller units. Here are the pros and cons of each type:

AC is the most useful current. It will power all the appliances, tools, lighting fixtures—everything you've come to expect from electricity. It is also much easier to transmit over long distances without big power losses. But an AC generator requires careful regulation, and this means a governor, which adds to the expense.

DC on the other hand has the advantage of lower expense because there is no need for a governor. But DC is not as convenient as AC. It won't power most appliances, TV or radio, fluorescent lighting and so on—at least not without rewiring the appliances or changing motors.

I personally see little point in spending $10,000 or more to produce DC current when our entire country is geared to run on AC. When you spend that kind of money you should expect to get power that closely approximates the stuff you get from your local utility.

Costs vs. Payback. Before we go on, now is a good time to think about the economics of water power. There are two ways to figure how long a water power unit will take to start breaking even with the rates you'd pay the power company for the same current. If you already have utility power you can simply divide the cost of your entire installation by your average monthly bill. This will tell you instantly how many months it will take to break even. If you spend $10,000 to install a system that provides you with all the power you used to get for, say, $50 a month, it will take you 200 months or about 16 years to break even—assuming electric rates don't go up. Of course they will, so your break-even date will come even sooner. And once you reach the break-even point your power is essentially free.

Very often, however, water power may be considered for a site that has no utility power. And if the nearest utility line is some distance away, you'll have to consider the costs of

WATTAGES FOR AC APPLIANCES

Name	Watts	Hrs/Mo.	KWHRS/Mo.
Airconditioner, central			620
Airconditioner, window	1,566	74	116
Blanket, electric	177	73	13
Broiler	1,436	6	8.5
Clock	1-10	continuous	1-4
Clothes dryer, electric heat	4,856	18	86
Clothes dryer, gas heat	325	18	6
Clothes washer, automatic	512	17.6	9
Clothes washer, wringer	275	15	4
Coffee pot		10	.9
Cooling, refrigeration	¾-1½ ton		200-500
Dehumidifier	300-500		50
Dishwasher	1,200	25	30
Disposal	445	6	3
Electrocuter, insect	5-250		1
Electric oven	3,000-7,000		100
Fan, attic	370	65	24
Fan, kitchen	250	30	8
Fan, 8-16 in.	35-210		4-10
Freezer, food, 5-30 cu. ft.	300-800		30-125
Freezer, frostless, 15 cu. ft.	440	330	145
Frying pan	1,196	12	15
Furnace, electric control	10-30		10
Furnace, oil burner	100-300		25-40
Furnace, blower	500-700		25-100
Furnace, stoker	250-600		3-60
Hair dryer	900-1,200	5	½-6
Hi-fi Stereo			9
Humidifier	500		5-15
Iron	1,088	11	12
Lawnmower	1,000	8	8
Light bulb, 75w	75	120	9
Light bulb, 25w	25	120	3
Projector	500	4	2
Radio, console	100-300		5-15
Radio, table	40-100		5-10
Range	12,200	8	98
Record player, transistor	60	50	3
Record player, tube	150	50	7.5
Recorder, tape	100	10	1
Refrigerator-freezer, 14 cu. ft.	326	290	95
Refrigerator-freezer, 14 cu. ft., frost free	615	250	152
Rotisserie	1,400	30	42
TV, b & w	237	110	25
TV, color	332	125	42
Toaster	1,146	2.6	3
Typewriter	30	15	.5
Vacuum cleaner	630	6.4	4
Water heater	4,474	89	400
Water pump (shallow)	½HP		5-20
Water pump (deep)	⅓HP		10-60

having the line run to your home. This can often amount to several thousand dollars. The charges vary from place to place, but when you begin to add a few thousand dollars to your costs of obtaining utility power, your own hydroelectric plant begins to look a lot better.

If your home is not yet built, has no power, or if you for any other reason have no past electric bills to compute your break-even date, you can use other techniques. One is to find out what the average monthly bill is in your area. Another is to estimate how much electricity you may use per month, and multiply this by the rate the local utility charges. To do this, look up the appliances you intend to run in the accompanying power consumption table. See what the consumption is in KWHRS per month for each appliance. Add all these figures together and multiply the total by your local cost per KWHR.

CONSTRUCTION OF THE HYDROELECTRIC SYSTEM

Once you've satisfied yourself that hydroelectric power makes sense for you, both in terms of economics and energy needs, you can plan for construction. First step should be to investigate the legal aspects of your installation. Check with local authorities on water rights. Recent environmental legislation aimed at protecting wetlands may be a problem. Be sure to get all legal problems out of the way before you begin work or you may end up wasting a lot of time and money.

Once the legal aspects are ironed out, you can begin planning in earnest. The major makers of small hydroelectric equipment are all very willing to help you plan your installation. The Leffel Company has a staff of engineers that can help you pick the right equipment and give advice on the best way to install it. They can even send an engineer to your site if required. In this respect they have a big advantage over European equipment makers. Since these services are available, it makes sense to put them to use right from the beginning; they'll help you avoid mistakes and false starts that can cost a lot of time and money.

Begin by filling out an information sheet like the one printed below. Most of the information is easy to provide. The only entry that may give you trouble is the area of the pond formed above the dam. If you already have a pond this is simple enough. If the pond is roughly rectangular, multiply length in feet by width in feet and divide by 43,560 (number of square feet per acre). If the pond is roughly round, multiply the square of the radius by pi (3.14) and divide by 43,560. If the pond is roughly oval multiply long diameter by short diameter by .79 and divide by 43,560.

No pond and no dam at present? Then you'll have a little extra work to do. In determining head, you proba-

bly already established a tentative location and height for your dam. This in turn gives you a tentative water level for the pond water. To determine the area of the hypothetical pond, all you have to do is plot a contour line on the land upstream from your dam. And this contour line should be at the same elevation as the top of your hypothetical dam spillway. Plotting the contour line is a job for a surveyor's transit level. You can buy or rent one. Buying isn't a bad idea since the tool will come in handy throughout construction of the site.

You'll get instructions with the transit, so we won't go into great detail on plotting the pond here, but essentially this is what you'll do:

Set up the transit at a convenient point upstream from the dam. Level the transit. Then have a helper hold a sighting pole at the position of the dam, with the base of the pole at the

---

What head can be developed?_____ feet

Average stream flow⎱
(cubic feet per minute)⎰ _____ minimum flow_____ maximum flow_____

What pondage will be formed above the dam?_____ acres

How high does tail water rise during floods?_____ feet

Is the site developed at present?_____

Give distance from dam to where plant will be installed_____ feet

Give distance from plant to point of power use_____ feet

What kilowatt capacity is desired?_____ Alternating Current_____
                                           or Direct Current_____

Volts_____ Phase (If AC) _____ Cycle (If AC)_____

Name_____ Date_____

Street_____ County_____

City_____ State_____ ZIP_____

Filling out an information sheet like this one will give your supplier the facts he needs to help you plan your hydroelectric power plant. A small sketch of your site is useful too.

level of the proposed spillway. Sight on the pole with the transit, and mark the pole wherever the horizontal cross hair intersects it. Now your helper can simply take the pole upstream while you follow him through the transit. Have him hold the pole at several points on the stream bank, moving up or down the bank until the point you marked on the pole matches up in your crosshairs. Drive a stake in the ground at the base of the pole for each point you plot, and make enough sightings to establish the outlines of the pond. Then measure the pond and calculate its area as described above.

DAMS

Dams can range in cost and simplicity from the earth dam all the way up to the concrete dam. No matter which type you choose, however, make sure you build it properly. A dam that washes out during high water can mean more than inconvenience and expense. It can cause severe damage, even death.

Dam Building Basics. Four types of simple dams shown here are all suitable for hydroelectric installations. Whenever you have a choice, however, avoid dams with wooden construction elements. In time the wood will

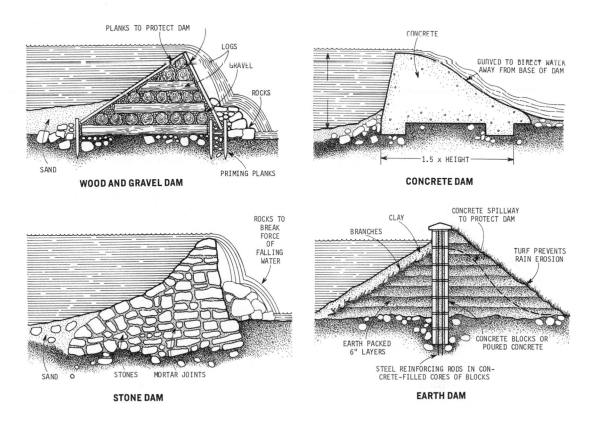

PLANKS TO PROTECT DAM
LOGS
GRAVEL
ROCKS
SAND
PRIMING PLANKS
WOOD AND GRAVEL DAM

CONCRETE
CURVED TO DIRECT WATER AWAY FROM BASE OF DAM
1.5 x HEIGHT
CONCRETE DAM

ROCKS TO BREAK FORCE OF FALLING WATER
SAND
STONES
MORTAR JOINTS
STONE DAM

CLAY
BRANCHES
CONCRETE SPILLWAY TO PROTECT DAM
TURF PREVENTS RAIN EROSION
EARTH PACKED 6" LAYERS
CONCRETE BLOCKS OR POURED CONCRETE
STEEL REINFORCING RODS IN CONCRETE-FILLED CORES OF BLOCKS
EARTH DAM

WOODEN CRIB DAM

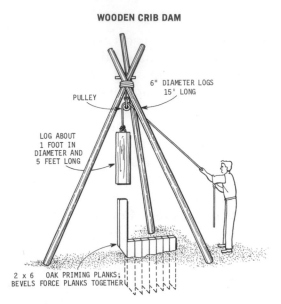

PULLEY

6" DIAMETER LOGS
15' LONG

LOG ABOUT
1 FOOT IN
DIAMETER AND
5 FEET LONG

2 x 6 OAK PRIMING PLANKS;
BEVELS FORCE PLANKS TOGETHER

Bevels on the "priming" planks ensure a tight seal when driven. Tripod legs move one at a time, allowing relatively easy adjustment of the strike zone of the log driver.

rot out. Cedar might last a lifetime, but other woods may let go in a matter of just a few years.

The key to building any dam is to make sure it doesn't allow seepage. Once seepage starts it will eventually wash a dam away. This very thing happened recently to a large earth dam in Idaho. Resulting damage was in the millions of dollars.

Your main seepage problems occur at the base of the dam. Water pressure is greatest there, and the junction between the dam and the ground it rests on provides a natural point for leakage. The method of sealing this joint will vary with the dam type and the nature of the substrate the dam rests upon. The wooden or crib dam rest-

ing on a clay or heavy soil base can be sealed with heavy planks driven into the base. The planks should be beveled as shown so they pack tightly against one another as they are driven. A simple log pile driver simplifies the setting of the "priming" planks. This same crib dam resting on a solid rock base would prevent a serious sealing problem. There's no way to drive planks into solid rock, so another dam type would be the best solution.

Concrete dams are sealed by keying as shown on the previous page and on the next page. A concrete dam is expensive, and despite its apparent strength, it's vulnerable to damage due to settling of the soil it rests on. It's worth the extra expense of hiring an engineer to help plan your dam if you decide on concrete. The same holds true of a stone dam.

You can allow water to spill directly over the top of a stone or concrete dam, but running water will quickly erode earth. The solution is to provide a spillway. This is a section of the dam lower than the main body of the dam. It should be lined with concrete or wood to prevent any erosion from starting. Any water flowing over the spillway of any dam should be directed away from the base of the dam to prevent the possibility of undermining the base.

Hardware. In planning your dam, don't forget to consider the problem of feeding the impounded water to your turbine. Leffel's Hoppes units

HYDROELECTRIC INSTALLATION

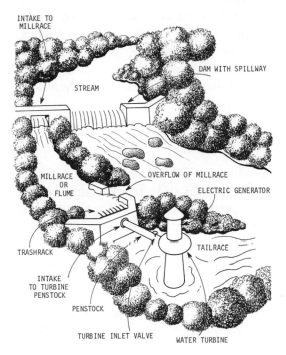

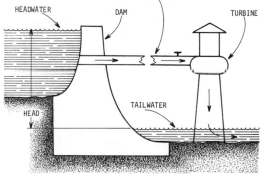

This hydroelectric installation with flume (millrace) illustrates the elements required for a site with little natural head. Water is impounded by the dam. Water required for generating power flows through the intake to the millrace and on to the generating plant. Excess water behind the dam flows over the spillway. Water flowing down the millrace is filtered by an iron grating called a trashrack, which removes sticks and other debris that could clog the turbine. Water passes from the millrace to the penstock and on through the turbine, in this case direct coupled to the generator. Spent water flows through tailrace and returns to the stream. Any excess water in the millrace passes over the millrace overflow and returns to the stream.

The ideal location for a hydroelectric plant is where a stream drops quickly at a waterfall. This gives you a lot of head very simply and cheaply. The inlet pipe is short, so cost is low and power loss in the pipe is kept to a minimum. When stream bed slopes gently you can get extra head by locating the power plant downstream some distance from the dam. Water from the dam can be carried via a pipe, or through an open channel called a millrace or flume, built up alongside the stream. Which way you decide to build is a matter of balancing costs. Remember that increasing head will reduce the cost of the generating equipment required to produce a given amount of power. On the other hand, costs of running a long pipe or flume may offset the savings you make on your generating equipment. One other factor to consider if flow is low: Increasing head will give you more power from a given amount of water.

are connected to an iron pipe which is usually embedded in the dam at its intake end. Embedding this pipe in concrete is a simple matter, and the resulting joint will remain tight. Concrete dams and those made of rock and concrete easily lend themselves to this type of installation. Earth and crib dams do not.

A hydroelectric installation includes a lot more than just a dam and a turbine. Even the simplest installation will also require a filter (called a trashrack) to keep debris out of the turbine. You'll also need a solid foundation to accept the turbine, and in some cases, several other features such as shutoff valves, extended penstocks, flumes, tailraces, and so on. These features are shown on the preceding page so that you can incorporate the features required for your particular situation.

Sources for More Information. Though some basic dam-building considerations are presented here, you shouldn't begin construction without acquainting yourself fully with the art of dam building. For starters, get a copy of *Small Earth Dams,* Circular 467. It's free from the California Agricultural Extension, 90 University Hall, University of California, Berkeley, CA 94720. Another source of information is *Ponds for Water Supply and Recreation* (Number 387) from the Superintendent of Documents, U.S. Government Printing Office, Washington, D.C. 20402. Price: $1.25. If these two booklets are inadequate for your needs, you can turn to *Design of Small Dams* ($12.65) also from the Government Printing Office. This book will prove more than adequate.

OTHER FORMS OF WATER POWER

So far, we've been talking about water power only from hydroelectric generating plants. What about the old fashioned water wheel? The water wheel was once an American fixture, but today it's an endangered species. The water wheel is primarily a slow-speed, low-torque power producer. Its versatility is limited. When you use water power to produce electricity, you have a form of energy that adapts easily to any job. The water wheel is at its best only when working at a slow, plodding chore such as grinding grain. Using a water wheel to produce electricity is possible, of course, but it requires a system of belts or gears to step up the speed, and it is difficult to regulate unless you'll settle for DC power.

In short, the water wheel is rare because it doesn't make economic sense. If you happen to come across one that's in actual working order, and you have a real need for low-speed power, fine. But there's not much point in building one from scratch because too much expense and labor are involved.

HYDRAULIC RAM

Here's a water-powered tool that does make sense. A hydraulic ram is about as close as you'll ever come to a perpetual motion machine. It can take the force of water falling a short distance (as little as 20 inches) and use it to pump a portion of that water to a height of 25 feet for each half-foot of fall. Read the description below, "How a Hydraulic Ram Works," to see how this neat bit of magic is performed.

There are two ways to get a hydraulic ram. You can make your own fairly simply, with a little help from a machine shop. VITA (Volunteers in Technical Assistance, 3706 Rhode Island Avenue, Mt. Rainier, MD 20822) sells a 25¢ booklet on building rams. Or you can buy a ram for prices running from around $300 to $3,000 from Rife Hydraulic Engine Manufacturing Company, Box 367, Millburn, NJ 07041. More about that later.

Buying a ram has some advantages over making your own. First, you'll get engineering assistance from the ram maker. You'll know how much water you can pump to any given height, and you'll know how much water will be required to do the pumping. Making your own ram is a little bit hit and miss. It is, however,

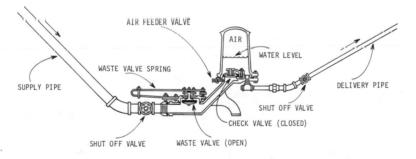

How a Hydraulic Ram Works. Water from the source flows down the supply pipe, and flows out of the ram through the waste valve. As the velocity of this falling water increases, the pressure closes the waste valve. This instantly stops the flow of water, but the inertia of the water creates tremendous pressure. This pressure opens the check valve, and water flows into the air chamber.

The water entering the air chamber quickly raises the pressure in the chamber. This forces the check valve shut. Now the pressure in the chamber forces water up the delivery pipe to its ultimate destination. Meanwhile, the pressure on the waste valve has dropped. The waste valve opens, water again starts to rush down the supply pipe and out the waste valve, and another cycle begins.

During each cycle, a small amount of air enters the air chamber through the air feeder valve. This valve must be adjusted so the right amount of air enters the chamber. Too much air, and the ram starts pumping only air. Too little air and the ram will lose its air cushion. The ram will begin to pound and may quickly break if the valve is not adjusted.

How fast does the ram run? It may cycle at a rate of anywhere from 20 to 100 strokes per minute. The slower it runs, the more water it consumes, and the more it pumps. You adjust the ram by changing the tension on the spring controlling the waste valve.

cheaper than buying, and if you find the ram you build doesn't pump enough water to suit your needs, you can always build another and run the pair in tandem. If you go this route, you can hook both rams to the same delivery pipe to save money. But each ram will require its own supply pipe.

Determining Available Power. Buying a ram is a lot like buying a hydroelectric plant. You can dig up a few basic facts about your situation and your requirements, send them off to Rife, and they'll tell you what you need to get the job done. See the illustration that is shown below for the infor-

mation they require. Measure the flow of your supply source and the vertical fall or head, as described earlier this chapter in the section on turbines. One bit of advice: Rams can run on much less water than turbines. If you have a low-volume source of water you can often measure its flow quite easily by directing it into a bucket of known volume. Just measure the time it takes to fill the bucket. If a two-gallon bucket fills in 15 seconds, flow is eight gallons per minute.

One more note: If you use the float or weir methods as described earlier, your results will read out in cubic feet per minute. A cubic foot contains about seven and a half gallons, so be sure to multiply your cubic foot readings by $7\frac{1}{2}$. (Rife asks for flow in gallons per minute.) Remember to make these measurements during late summer, or whenever the flow of your water source is at its lowest.

Once you've filled out an information sheet, you can quickly calculate, on your own, the maximum volume of water deliverable by a ram. This simple formula does the trick:

$$\frac{V \times F}{E} \times \frac{3}{5} = D$$

In this formula, V is the volume of flow available to you, in gallons per minute. F is the vertical fall in feet, and E is the elevation to which the water will be pumped (also in feet). The $\frac{3}{5}$ is a constant reflecting the ef-

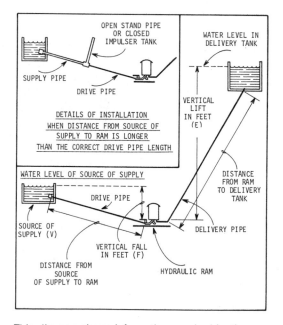

This diagram shows information required by the ram maker before he can advise you on selection and installation of a ram.

ficiency of the ram. (This can be lower with home made rams.) Your final result, D, is the volume of water delivered in gallons per minute.

If you're interested in computing delivery in gallons per hour, multiply D by 60. Gallons per day? Multiply D by 1440. Note: This formula gives you the absolute maximum amount of water that can be produced by using all the water available at your source. In reality, however, it's best not to use all of your water source; this doesn't leave any insurance water in times of low flow. And if a ram consumes water at a rate faster than your spring or other source can supply water, the ram will stop.

Basics of Ram Installations. Rams are relatively simple to install, especially if your conditions are ideal. In some situations you'll have to add a few features to your installation that won't be required for the ideal situation. We'll cover those later, but for now, let's go over the basic installation.

Begin by deciding where to put the ram. A number of factors will influence this decision. First of all, placing the ram at a low elevation will give you more fall. More fall means more water can be pumped, and smaller, cheaper rams can be used. So you'll want to place the ram where it will provide as much fall as possible.

But that's not all you have to consider. The drive pipe from your source of water to the ram should be the correct length. For rams with a fall of up to 15 feet, the drive pipe

should be about five or six times as long as the fall. For rams with a fall up to 25 feet, the pipe should be about four times the fall. With 50-foot falls, the drive pipe should be about three times as long as the fall.

Finally, you should place the ram as close to the source of supply, and the delivery point, as possible. This will keep pipe costs down, and also limit friction within the pipes. Less friction means greater delivery.

The ram must be placed on a solid foundation. A concrete slab or a large flat rock is ideal. The slab should slope away from the ram's waste valve, so excess water flowing from this valve will drain off. After the ram is set up and running, it's a good idea to surround the ram with a concrete pit high enough to protect the ram from flood waters. A drain tile allows the ram's waste water to flow out of the pit.

The drive pipe from source of water to ram should be as straight as possible. It should be firmly anchored, and preferably buried underground. But don't bury it until you run the ram and check the pipe for leaks. The intake end of the pipe should be protected with a strainer to keep out debris that could clog the ram. A gate valve in the drive pipe— just above the ram—will let you close off the line whenever you have to work on the ram.

The delivery pipe—like the drive pipe—should be as straight as possible to reduce friction. If turns are required to conform with the terrain,

As described in the accompanying text, here are details of a ram installation designed to collect water for domestic use. Note that water from storage is gravity-fed to the house.

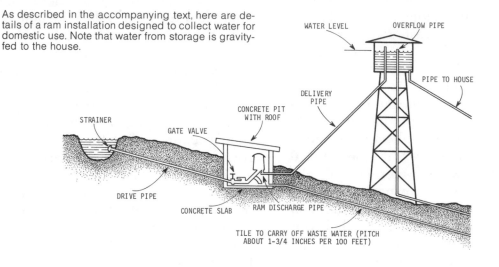

make the turns as gentle as possible. Two 45-degree angle fittings are better than a single 90-degree elbow. The delivery pipe is best buried below frost in very cold climates. Generally, though, the pipe shouldn't freeze since water in it is always flowing.

A storage tank or holding pond will be required to hold the water once it has been pumped to its destination. Water for domestic use is best stored in a tank located above the level of the home. A pipe from the tank can then serve your home with water by gravity feed. Since the ram runs continuously, you'll have to provide some means of getting rid of excess water. Best way to do this is to run an overflow pipe down through the bot-

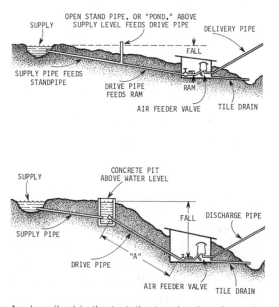

As described in the text, the top drawing shows the placement of the stand pipe (an auxiliary pond) needed when the drive pipe is long and straight. For a steep drop, as in the bottom drawing, a pit serves as the angle connection.

tom of the tank. The level of the top of this pipe will determine the level of the water in the tank.

Problem Installations. Not all rams can be installed so that they are fed by a short, straight drive pipe. Special techniques may be required to adapt rams to different situations.

When you are forced to locate the ram so far away from the source (in order to get enough fall) that the drive pipe becomes longer than recommended, use the setup shown in the middle drawing. Here a supply pipe one size larger than the drive pipe feeds water to the drive pipe. At the junction you install an open-ended stand pipe, or "pond," made of pipe two sizes larger than the supply pipe. The supply pipe feeds the stand pipe, and the stand pipe feeds the drive pipe. This technique is for situations where supply and drive pipe can make a continuous straight run.

When supply and drive pipe cannot make a straight run, use the setup in the bottom drawing. Install the upper end of drive pipe at least one foot below water level to avoid whirlpooling that may let air into the pipe.

Special-purpose Rams. Rife makes a double-acting ram to solve a special problem. Say you have a very limited amount of potable water—enough to serve your living needs, but not enough to power a ram. If you also have a larger source of nonpotable water you can use this source to run the ram which will then pump the potable water. This double-acting ram is just the ticket for pumping water from a small spring—as long as you have another source of water to run the ram.

BIOFUELS

No doubt you've heard stories about the man who runs his car on methane produced from chicken manure. This kind of chemical magic is possible. But it's just not practical unless you've got a lot of barnyard animals supplying the wastes. The same is true of methanol production. The production of fuel gas from wood, however, does hold promise for the average homeowner.

Let's first take a look at the state-of-the-art for producing fuel from wastes.

METHANE

Turning organic wastes into methane gas may not yet have the alchemical appeal of turning lead into gold, but in the future, things could change. After all, the world has more gold than it can profitably put to use. Energy, on the other hand is in short supply, and it's the driving force of the world. All told, the benefits of turning wastes into energy are staggering. Unfortunately, to make the process economically justifiable, it has to be carried out on a large scale. And that leaves the homeowner, the homesteader, and even the small farmer out of the picture. More about that later, though. First let's take a look at some of the benefits of generating methane from wastes.

Obviously, the primary benefit is

METHANE FLOW CHART

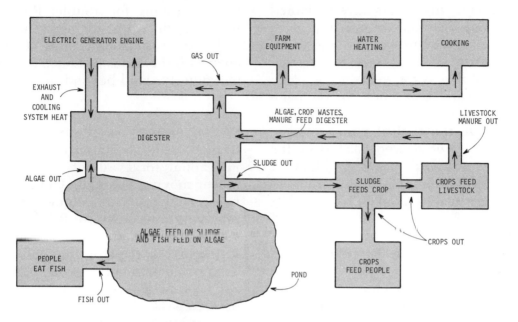

This idealized flow chart shows the self-sufficient nature of methane generation. This digester feeds on algae from a culture pond and on animal and crop wastes. The gas produced heats water, cooks food, and runs an engine driving a electric generator. Waste heat from the engine's exhaust and cooling system is recycled to the digester to keep it the ideal temperature of 95 degrees. And this stimulates gas production. Note that the engine running the generator can heat water for warming the digester almost as efficiently as a water heater, and also it produces electricity at the same time.

Nutrient-rich sludge from the digester is tapped off and used to feed crops and to feed algae in the culture pond. Fish in the pond feed on algae and produce food for humans. Excess algae is recycled to the digester. Crops grown on sludge feed people and livestock. Crop wastes go back into the digester. So do manures produced by animals eating sludge-fed crops.

The process at first seems to be a perpetual motion phenomenon, violating all the laws of science. Actually, the process is fueled by solar energy captured photosynthetically by the algae and crops.

methane. It is produced by the decay of organic wastes such as manure and shredded vegetation. The decay is anaerobic, that is, brought about by bacteria that live only in the absence of oxygen. With manure from 40 head of cattle, a farmer could produce enough methane to satisfy the energy requirements of his entire farm.

That's not all. A by-product of

methane production is a liquid sludge with extremely high value as a fertilizer.

Together, the methane and sludge are a boon to the farmer. They can help free him from his almost total dependence on petroleum-derived fuels and fertilizers. As Barry Commoner notes in his *The Poverty of Power,* this dependence has raised the farmer's operating costs year after year, while his profits have remained the same. Methane generation gives the farmer a source of fuel and fertilizer at reduced cost, and at a cost that will not rise as the years go by.

But again, that's not all. Methane production also solves other problems. On the typical farm of today, manure is *not* viewed as an asset. It's simply a big disposal problem. A great deal of labor is required just to keep the stuff under control. In addition, it releases nutrients that run off with the rain and pollute our waters.

Using the manure for methane production can solve these problems. John Fry—who ran a huge pig farm in South Africa—once built a methane digester for about $10,000. He figures the benefits this way:

"The gas alone must have paid for my installation in three or four years. I would say the digestor saved me far more than that in labor . . . and an even greater (amount) again in the fertilizing effluent. . . ."

Scale, Hazards, Local Codes. Okay, that's the rosy side of the picture. Now for the dark side. Efficient, low-cost methane production can only be carried out on a large scale. The initial cost of a generator is high. Of course, many small units have been built from junk parts, but in the interest of safety, this isn't the best way to go. Methane can be very explosive, and pumping the stuff around in gar-den hoses and rusty old pipes and tanks is asking for trouble. Furthermore, if you live in an area regulated by building and health codes, compliance with the codes can push the cost of a digester well beyond reason.

Methane Applications. Putting the gas to good use is a problem, too. You can just about forget about using it to power moving vehicles. Unlike propane, methane won't liquefy under relatively low pressure. By the time you burn enough fuel to run compressors to liquefy the stuff, you will have used as much energy as you will ever get back when you burn the compressed gas.

Of course you can store the stuff in larger tanks under less pressure. But a cylindrical tank five feet long and nine inches in diameter will hold the equivalent of only about 3½ gallons of gasoline.

Furthermore, it's difficult to generate a useful amount of gas. Small-scale homestead digesters with tanks in the 600-gallon range won't do the job. For example, Mother Earth News built a digester with a tank measuring four feet in diameter and nine feet in height. That works out to about 113 cubic feet or 850 gallons. The digester produced 41 cubic feet of gas per day. Mother Earth claims that since there wasn't enough storage capacity (there's that storage problem again) more than 41 cubic feet a day were

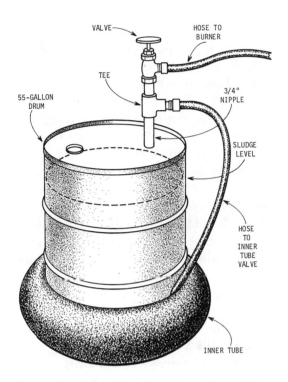

VALVE

HOSE TO BURNER

TEE

55-GALLON DRUM

3/4" NIPPLE

SLUDGE LEVEL

HOSE TO INNER TUBE VALVE

INNER TUBE

An experimental methane digester can be made from a 55-gallon drum, some plumbing supplies, and an old truck or tractor inner tube. To put it into operation, pour about 40 gallons of slurry into the drum. (Slurry is a creamlike mix of organic wastes and water.) By using only about 40 gallons of slurry, you allow room at the top of the tank. Before hooking up the inner tube, purge it of all air. For fastest results, keep the drum at 95 degrees. Temperatures as low as 80 will suffice but will result in slow, low-volume gas production.

Gas will begin to form and inflate the inner tube in a few weeks or so. For safety, do not burn the first tubeful of gas. Open the valve and press the inner tube flat to expel as much gas as possible. Then close the valve and let it refill. The gas is now safe to burn. Failure to follow this procedure may result in a methane/air mix inside piping and the inner tube. If ignited, the gas could flash back and explode.

When gas production stops (after a few months) empty all but a few gallons of the slurry on garden crops and refill the drum. By saving a few gallons of old slurry in the drum, you provide a bacteria culture for the next batch.

produced, but only 41 cubic feet could be retained.

For the sake of this argument we'll assume the generator produced a total of 50 cubic feet per day. Now 50 cubic feet of bio-gas is not a very useful quantity. It can run a typical gas refrigerator for about half a day. Or it can cook simple meals for a family of four for only a day.

Costs vs. Payback. The costs of building a generator of this capacity vary. But a price around $800 is close to what early digester builders have had to pay, even using a lot of junk parts. How long would it take for such a generator to pay off? Well, 50 cubic feet of natural gas at a price of $3 per 1,000 cubic feet would be worth a total of 15 cents. If bio-gas had the same fuel value as natural gas, our 50 cubic feet of bio-gas would also be worth 15 cents. In reality, bio-gas has a fuel value equal to only about two-thirds that of natural gas. So figure your day's output of bio-gas is worth a dime. Divide a dime into $800 and you get 8,000 days till payoff. That's roughly 22 years.

So that's the not-too-promising picture on building a small-scale methane digester. Payoff time is 22 years, compared to a matter of four or so years for an installation the size of John Fry's. Not only that, the amount of gas you get from a small unit is just about enough to cook your meals, no more.

The Cost-effective Plant.   Then, how big do you have to go to make a methane generator economically viable? A minimum size might be a plant capable of processing the wastes from about 40 cows. If the plant were built cheaply enough, the baseline size might drop to as low as 20 cows. Even a 20-cow plant is out of the reach of most farmers and homesteaders. Cornell University estimates the costs of such a plant on a typical dairy farm might run around $20,000. To handle the wastes of 20 cows—allowing for about 8½ gallons per cow per day— you'd need a tank with a capacity of around 12,500 gallons. That's a tank 25 feet long, 10 feet wide, and 6 feet deep. John Fry's pig farm plant had two tanks, each twice this size. An undertaking of this magnitude is too much for most farmers to handle without expert engineering advice.

The real promise of methane lies in large-scale generating plants. These could produce methane on a grand scale, using agricultural wastes, or vegetable matter produced specifically for the purpose of making gas. High on the list of vegetable-gas sources are water hyacinths, algae, kelp, and kenaf. Effluent from the digester could be recycled to the growing vegetation, increasing its growth rate. The gas could be burned to generate electricity. The waste heat from the engine turning the generator would keep the temperature of the methane digesters at the right level. This would increase gas production, and the excess gas could be used to take the place of dwindling supplies of natural gas.

Dr. Donald Klass of the Institute of Gas Technology estimates that such a process could provide us with gas to meet 65 percent of our present needs.

But all of this is in the future, and it's going to be in the hands of giant commercial interests. Is there anything you can do to produce methane yourself? There's no harm in trying your hand at methane generation on a small scale if you can keep your costs and your expectations down. A few small generators may be worth the bother even if you don't use the gas. The effluent is valuable in itself, better than the compost you'd get if you let wastes decay aerobically in a heap.

Sources for Methane Information.   If you're interested in small-scale methane digesters, here a few publications to check out:

- *Methane Digesters for Fuel Gas and Fertilizer,* from the New Alchemy Institute, Box 432, Woods Hole, MA 02543.

- *A Homesite Power Unit: Methane Generator,* by Les Auerbach, 242 Copse Road, Madison, CT 06443.

- *The Mother Earth News Handbook of Homemade Power,* from Bantam Books, 666 Fifth Avenue, New York, NY 10019.

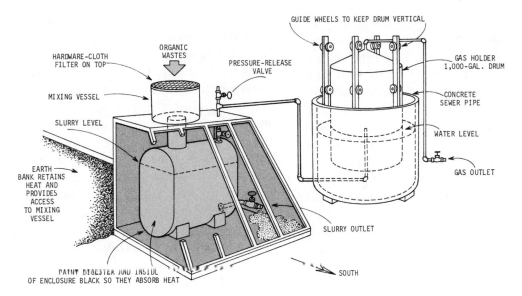

GUIDE WHEELS TO KEEP DRUM VERTICAL

HARDWARE-CLOTH
FILTER ON TOP

ORGANIC
WASTES

PRESSURE-RELEASE
VALVE

GAS HOLDER
1,000-GAL. DRUM

MIXING VESSEL

CONCRETE
SEWER PIPE

SLURRY LEVEL

WATER LEVEL

EARTH
BANK RETAINS
HEAT AND
PROVIDES
ACCESS
TO MIXING
VESSEL

GAS OUTLET

SLURRY OUTLET

PAINT DIGESTER AND INSIDE
OF ENCLOSURE BLACK SO THEY ABSORB HEAT

SOUTH

A homemade methane digester made from an old fuel-oil tank might look like this. It's housed in a glass or plastic-covered lean-to enclosure that uses the sun to help maintain the necessary temperature. You can cut down on sludge accumulation in the tank by fitting the mixing vessel with a filter made of hardware cloth. This keeps out straw and other large chunks of indigestible material. In addition, the outlet of the mixing vessel should be above the bottom of the vessel so grit and sand have a chance to settle out.

The gas collector itself consists of a dome floating in a bed of water. Gas from the digester bubbles up into the dome; the water provides a gas-tight seal. Weight of the dome pressing down on the cushion of gas provides the necessary pressure to move the gas through the outlet. A good pressure for moving gas is about eight inches of water, or 41 pounds per square foot. So to provide the proper pressure, the weight of the dome divided by the area of its cross section (Pi times radius squared) should equal about 41. The gas dome can be weighted or counterweighted to achieve the optimum pressure.

To increase the capacity of a setup like this, you could hook two or more digesters to the gas storage dome. While one of the batch-fed digesters is out of operation the others will keep producing gas.

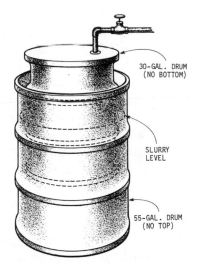

30-GAL. DRUM
(NO BOTTOM)

SLURRY
LEVEL

55-GAL. DRUM
(NO TOP)

Even simpler methane digesters combine gas storage and production in one vessel. Instead of floating the gas dome on a bed of water, you float it on a bed of organic slurry. The gas produced rises directly into the dome for storage. A digester of this type is easy to make, simply by using a 55-gallon drum and a 30-gallon drum. Cut the top out of the big drum, and the bottom out of the small one. Fill the 55-gallon drum about three-fourths full of slurry. Then place the open-bottomed, 30-gallon drum into the large drum. Press it all the way down into the drum. When gas forms, it will collect in the small drum.

METHANOL

If methane production is out of the picture for most of us, what about methanol? This is another fuel that can be produced from organic wastes including rubbish, wood scrap, and farming wastes. It makes an excellent fuel in automobiles with a few minor modifications to the car. Running on pure methanol requires intake manifold heating, rejetting of the carburetor, a larger gas tank (to compensate for methanol's lower heat of combustion) and a starting aid (to compensate for the difficulty of cold-starting alcohol-power engines). In return your car will give you: "Improved economy, lower exhaust temperatures, lower emissions, improved performance," according to doctors Thomas Reed and R. M. Lerner of MIT.

Additionally, methanol is an easy fuel to handle in contrast to methane, hydrogen, natural gas and other proposed alternate fuels. It could be incorporated into our national network of gas stations without much problem. Unfortunately, there's no way to produce the stuff on a homeowner level. So production is once again in the hands of the big energy producers. At present, production is very limited—about one percent of the total for gasoline production. And the methanol being produced is made from natural gas—one of our scarcest nonrenewable energy sources.

Need for National Program. Producing useful fuels such as methane and methanol from renewable sources of organic materials obviously makes a lot of sense in an energy-hungry country like the U.S. But very little work of this kind is being done. Nor is it even being contemplated for the near future. Since the manufacture of these fuels is beyond the scope of the homeowner, there's not much you can do yourself to change things. A change in national policy favoring these fuels will come about some day, but probably only when the need has become painfully obvious. About the only way you can personally involve yourself in methane or methanol production is to work for it on the political level, through lobbying.

WOOD GAS

What about wood gas as a fuel? Not many Americans are aware of the fact, but it's possible to run a car on wood, charcoal, or coal. And in theory, it's a simple process. Your fuel—let's say wood—is burned in a sort of airtight stove, with an insufficient supply of oxygen. The resulting exhaust gas from the stove is largely carbon monoxide. The carbon monoxide is fed to the car's engine, where it is burned to form carbon dioxide. The process works. Thousands of European cars, trucks and tractors were fitted with these little gas factories (called gasogens) during World War II.

Drawbacks? Plenty. First of all, performance is reduced considerably. And regulation of the gasogen is

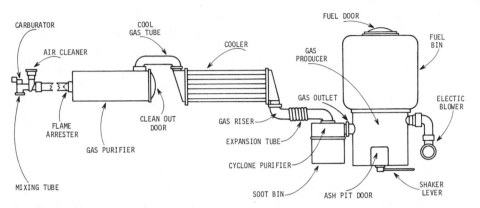

This charcoal fueled gasogen is designed by System Svedlund of Sweden. In practice, the heart of the unit (vertical section at right of drawing) is usually mounted on the car's rear bumper. It extends high enough to somewhat block the driver's view to the rear. Charcoal gasogens run on fairly clean fuel compared to wood or soft coal versions. So the filtering and purifying system shown here is simpler than the one required for wood.

The electric blower causes the initial draft through the fuel bed. The fire is started with special matches.

Once the car is running, intake vacuum provides the necessary draft. Gas exits the generator at around 550 degrees F. Much of the soot is removed in the purifier and settles into the soot bin. The cooler lowers the gas temperature to increase its density and thus its Btu-per-volume value. This helps raise power already drastically reduced because the gas is diluted with air (mostly nitrogen).

The engine is regulated via the throttle which operates a butterfly valve (similar to a damper) located in the mixing tube.

The heart of the gasogen is the combustion chamber, or gas producer. It's like an airtight stove in which wood chips or other fuel can be burned with a carefully regulated air supply. With the air supply choked down, fuel burns incompletely to form carbon monoxide (CO). The CO is then piped to an engine where its burn is completed to form CO_2, or carbon dioxide. In essence the fuel is half burned in the gasogen in order to convert it to a gas that can then be burned the rest of the way in the car engine.

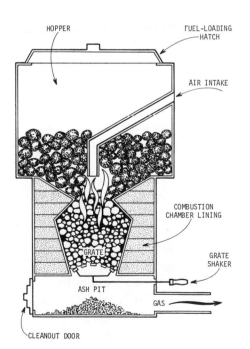

tricky. Wood and soft coal present special problems. The gas they produce is so full of tars and other engine-wrecking chemicals that filtration is a must. So is frequent filter maintenance. In addition, the gas from the gasogen must be cooled before it reaches the engine, or performance suffers even more.

So now the relatively simple gasogen is complicated with the addition of filter chambers, cooling tubes and so on.

Though no commercially-made gasogens are being manufactured at present, Sweden has been seriously considering the idea. Both Saab and Volvo have been running experiments with wood-chip-fueled gasogens. Current Swedish estimates place the cost of a mass-produced gasogen system at around $1,000. Further, the Swedes figure they could be into production of gasogens within six months in the event of a severe fuel crisis.

Build your own gasogen? It's possible, although you'd have to do a lot of experimenting and tinkering to arrive at a smooth-running product. The accompanying drawings show some of the main design features of past gasogens, and captions give the basics of operation.

Costs vs. Payback. If you do build your own, what can you expect in the way of performance and economics? Past experience with gasogens has revealed that power output is just about half what you'd get with gasoline as a fuel. As for economics, it all depends on what you have to pay for wood and for the gasogen. For example if you have a supply of free wood from your own land, and if you spend $500 to build a gasogen, your gasogen will pay for itself as soon as you drive your vehicle the same distance it would go on $500 worth of gasoline. If your car gets 20 mpg on gasoline, and gasoline runs 60 cents a gallon, your payoff comes at 16,666 miles. This isn't bad, and it would be even better if the price of gasoline rises, or if there is no gasoline available.

On the other hand, payoff comes later if you have to pay for your wood. The kind of mileage you get from wood can vary a great deal, but for calculations you can use the following figures: Good hardwood yields

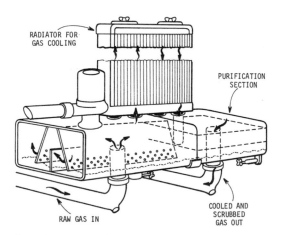

A cooler and purifier system must be added to the gas generator in order to make the gas palatable to an internal combustion engine. Uncleaned gas contains a variety of tars and acids, and is so hot that it must first be cooled before it will be dense enough for burning.

about one mile per pound of wood, or about 4,000 miles per cord. If you pay $60 per cord, you're paying the equivalent of 30 cents a gallon for gasoline, again comparing a gasoline mileage of 20 mpg.

Realistically, it wouldn't make sense for you to use gasogens unless you have an abundance of wood, and an extreme shortage of gasoline. For example, if you live in a remote area, far from a source of gasoline, and you need a fuel to run a gasoline engine (for generating electricity, powering a buzz saw or sawmill, or maybe a tractor) a gasogen might be a solution.

Sources for More Information. If you are interested in trying to design your own gasogen, here are some references that will help get you started.

Unfortunately, most are from technical journals published back in the 40s (when wartime fuel shortages spurred gasogen interest). They may be difficult to locate.

- "Thermodynamics of Producer Gas Combustion," July, 1945 issue of *Industrial and Engineering Chemistry,* Vol. 37, No. 7. Nine pages.

- *Gasogens,* Report Number 1463 from the Forest Products Laboratory, U.S. Forest Service, PO Box 5130, Madison, WI 53705.

- "Can We Use Wood to Beat the Gasoline Shortage?" *Popular Science* Magazine, January, 1944.

- "How to Run Your Car on Wood," *Mother Earth News,* No. 27, 105 Stoney Mt. Rd., Hendersonville, NC 28739.

INDEX